Praise for the series

'. . . allows a space for distinguished thinkers to write about their passions.'
The Philosophers' Magazine

'. . . deserve high praise.'
Boyd Tonkin, *The Independent* (UK)

'This is clearly an important series. I look forward to reading future volumes.'
Frank Kermode, author of *Shakespeare's Language*

'. . . both rigorous and accessible.'
Humanist News

'. . . the series looks superb.'
Quentin Skinner

'. . . an excellent and beautiful series.'
Ben Rogers, author of *A.J. Ayer: A Life*

'Routledge's *Thinking in Action* series is the theory junkie's answer to the eminently pocketable Penguin 60s series.'
Mute Magazine (UK)

'Routledge's new series, *Thinking in Action*, brings philosophers to our aid . . .'
The Evening Standard (UK)

'. . . a welcome new series by Routledge.'
Bulletin of Science, Technology and Society (Can)

Praise for the series

JOHN COTTINGHAM

On
the Meaning of Life

Routledge
Taylor & Francis Group

LONDON AND NEW YORK

First published 2003
by Routledge
2 Park Square, Milton Park, Abingdon, Oxon OX14 4RN

Simultaneously published in the USA and Canada
by Routledge
270 Madison Avenue, New York, NY 10016

Reprinted 2004, 2006

Routledge is an imprint of the Taylor & Francis Group

Typeset in Joanna MT by
RefineCatch Limited, Bungay, Suffolk
Printed and bound in Great Britain by
Biddles Ltd, King's Lynn, Norfolk

British Library Cataloguing in Publication Data
A catalogue record for this book is available from the British Library

Library of Congress Cataloging in Publication Data
A catalog record for this book has been requested

ISBN 0–415–24799–3 (hbk)
ISBN 0–415–24800–0 (pbk)

For
MLC

Preface

Questions about the meaning of life are closely intertwined with religious questions, and so there is an automatic risk of giving offence. Many religious adherents may be put off if the answers offered do not start from the doctrines they hold as central to any account of life and its meaning. Many atheists, by contrast, may be irritated that religious ideas should be allowed to intrude at all into our human struggle to find meaning in our lives. While I dare not presume to have avoided giving such annoyance on either side, it is my hope that by sidestepping a dogmatic stand-off on matters that may be beyond the horizon of rationally determinable knowledge, we can find space for a productive inquiry into what may broadly be called the 'spiritual' dimension of the quest for meaning. Among the book's eventual aims are to disclose something of the importance and preciousness of that dimension, to reveal how it connects with values and commitments that we all share, and to find a way of accommodating it without the sacrifice of scientific or philosophical integrity. My strategy is (deliberately) a gradual one, so I have to request the reader's patience if the goals are not placed fully in view until the final chapter; since the book is a short one, and aims to avoid technicality and philosophical jargon, I hope this request will not be too burdensome.

An earlier version of some portions of Chapter Three was presented at a conference on Spirituality at the University of St Andrews in March 2001, and I should like to thank the participants for much stimulating discussion. I am very grateful to Max de Gaynesford and to Jim Stone, who were kind enough to read the whole manuscript before publication and to make many acute and helpful suggestions and comments. I should also like to record my gratitude to my colleagues in the Philosophy Department at the University of Reading, whose friendship and support over the years has been a great source of strength. I am also greatly indebted to the administrators of the University's Research Endowment Trust Fund for a grant that provided an important measure of teaching relief while this book was being completed. My greatest debt is to my immediate family, for sharing with me their reflections on many of the themes found in this book, and for enriching my life in more ways than could ever be put into words.

J. C.
Reading, England
April 2002

The Question

One

'Alright', said Deep Thought. 'The Answer to the Great Question . . .'
'Yes!'
'Of Life the Universe and Everything . . .' said Deep Thought.
'Yes!'
'Is . . .' said Deep Thought, and paused.
'Yes!'
'Is . . .'
'Yes . . . !!! . . . ?'
'Forty-two', said Deep Thought, with infinite majesty and calm . . .
It was a long time before anyone spoke.
'Forty-two!' yelled Loonquawl. 'Is that all you've got to show for seven
and a half million years' work?'
'I checked it very thoroughly,' said the computer . . . 'I think the
problem, to be quite honest with you, is that you've never actually known
what the question is.'
'But it was the Great Question! The Ultimate Question of Life, the
Universe and Everything' howled Loonquawl.
'Yes', said Deep Thought with the air of one who suffers fools gladly,
'but what actually *is* it?'
A slow stupefied silence crept over the men as they stared at the
computer and then at each other.
'Well, you know, it's just Everything . . . Everything . . .' offered
Phouchg weakly.

Douglas Adams, **The Hitchhiker's Guide to the Galaxy**[1]

THE QUESTION THAT WON'T GO AWAY

Not all important-sounding questions make sense. For a fair
part of the twentieth century it was common in much of the
anglophone world to dismiss many of the traditional grand

questions of philosophy as pseudo-questions. People who felt perplexed by the ancient puzzle of the meaning of life were firmly reminded that meaning was a notion properly confined to the arena of language: words or sentences or propositions could be said to have meaning, but not objects or events in the world, like the lives of trees, or lobsters, or humans. So the very idea that philosophy could inquire into the meaning of life was taken as a sign of conceptual confusion. The solution to the problem, as Ludwig Wittgenstein once remarked, would lie in its disappearance.[2]

But somehow the problem does not go away; the search for life's meaning, confused or not, retains as powerful a hold on us as ever. The characters in Douglas Adams' Hitchhiker saga may seem absurd in their faith that a supercomputer could wrap it up for them, and hopelessly vague about how to formulate the problem in the first place, but a strong sense remains that the ancient quest that has held so many in thrall is more than just a philosopher's muddle.

For our human existence is mysterious – something strange, frightening, to be wondered at. Philosophy, said Aristotle, is the child of wonder;[3] and the capacity to be disturbed by what is ordinarily taken for granted is the hallmark of that questioning spirit that is inseparable from human nature itself. The human being is unique in that, as Heidegger put it, it is an entity for whom its own being is an *issue*. Or again: 'Man alone of all beings, when addressed by the voice of Being, experiences the marvel of all marvels: that what-is *is*.'[4]

What are we really asking when we ask about the meaning of life? Partly, it seems, we are asking about our relationship with the rest of the universe – who we are and how we came to be here. One aspect of this is a scientific question about our origins. To which the answer, only recently discovered, is breathtaking: we came from the stars. If we manage (the experience is rarer and harder now) to find a spot far away from the city, where no seepage of noise and dazzle pollutes the night, and look up in wonder at the vast and silent blackness of space from which numberless brilliant points of light shine down upon us, then what we see is the same material from which we, and everything else on this fragile planet, were once formed. We humans are part of the cosmos: not just as a pebble is part of a miscellaneous heap, not just as an item on a haphazard inventory that happens to include whatever the universe contains; but truly one with it, sharing its common origin, built of its stuff. We are formed of stardust.

Of it, yet alienated from it? It may be so. The ancient Stoics thought that our human rationality was a microcosm of a governing principle of Reason, the spiritual substance pervading the whole cosmos; centuries later, the rationalist philosopher Leibniz declared that 'there is nothing waste, nothing sterile, nothing dead in the universe'.[5] But the dominant view nowadays is that life and rationality are, cosmically speaking, local and untypical features of reality: nature is predominantly blind, irrational, dead. As the poet A. E. Housman lamented:

> . . . nature, heartless, witness nature,
>   Will neither care nor know
> What stranger's feet may cross the meadow
>   And trespass there, and go.
> Nor ask amid the dews of morning
>   If they are mine, or no.[6]

We humans may pride ourselves on our intellectual and cultural achievements, but against the backdrop of unimaginable aeons of time through which clouds of incandescent hydrogen expand without limit, we are a strange temporary accident, no more significant than a slime or mould that forms for a few years or decades on a barren rockface and then is seen no more.

Assessments of this kind may seem linked to a modern scientific understanding of our origins, but in an important sense they plainly go beyond science: they do not just report the 'facts', but talk about what those supposed facts 'mean' for us, for our sense of ourselves and our self-worth. And it is hard to see how such judgements about the significance of our lives can be established by scientific inquiry alone. To quote Wittgenstein again, this time sounding rather more hospitable to our grand question, 'we feel that even when all possible scientific problems have been answered, the problems of life have not been put to rest'.[7] Why exactly should this be so?

SCIENCE AND MEANING

Science has advanced so spectacularly and with such an accelerated pace in the last century or so that we may be tempted to suppose that given a bit longer it could even succeed in explaining why we are here and what our existence means. This appears to be the view of one of our most distinguished contemporary scientists, Stephen Hawking:

> Up to now, most scientists have been too occupied with the development of new theories that describe *what* the universe is to ask the question *why*... However, if we discover a complete [and unified] theory [combining quantum physics with general relativity]... we shall all... be able to take part in the discussion of the question of why it is that we and the universe exist. If we find the answer to that, it would be the ultimate triumph of human reason...[8]

The distinction between *what* something is and *why* it is has become something of a cliché; in similar vein, people often say science tackles *how* questions but not *why* questions. But in fact the distinction is not particularly helpful in sorting out what scientists characteristically do. Aristotle was rather more perspicuous in distinguishing four types of answer relevant to scientific inquiry:

(1) Answers indicating the component materials of which an object is made (its 'material' cause);
(2) Answers specifying the essence or kind of thing it is (its 'formal' cause);
(3) Answers pointing to the motive force that got it into its present state (its 'efficient' cause); and
(4) Answers citing the end or goal towards which it tends (its 'final' cause).[9]

Explanations of all four kinds can be good scientific answers to the question 'why?'

(1) 'Why was the bridge strong?' 'Because [material] it was made of steel.'
(2) 'Why do you classify that ice cube as water?' 'Because [formal] it is frozen H_2O.'
(3) 'Why did the billiard ball move?' 'Because [efficient] it was struck with a cue.'
(4) 'Why do trees have roots?' 'Because [final] in order to grow they need to take up water and nutrients.'

The last type of answer was particularly important in Aristotle's work, since he maintained that all things tend towards some natural end-state; but although modern scientists, especially biologists, still frequently use such goal-related or teleological explanations of phenomena, it has been a guiding principle since the seventeenth century that such teleology must always eventually be explained in terms of underlying microstructures of an entirely mechanical nature. It is in this sense that the great seventeenth-century philosopher-scientist René Descartes is often said to have banished teleology from science. He envisaged a unified style of explanation based ultimately on the universal laws of mathematical physics that governed the behaviour of all natural phenomena, celestial and terrestrial alike. There was no room for any irreducible purposiveness or goal-seeking deep down in nature. The job of the scientist was to subsume all observable events under the relevant mathematical covering laws; and in respect of these ultimate laws there was no attainable answer to the question 'why?' One could say – and Descartes did say – that God had decreed that it should be so; but he immediately added that the rationale for God's decrees was not for human scientists to discover: it was 'forever locked up in the inscrutable abyss of His wisdom'.[10] David Hume, writing a century after Descartes, took an essentially parallel line, though couched in entirely secular language: the job of science was to map the observable natural world, but any supposed 'ultimate springs and principles' of nature were beyond human power to fathom.[11]

Although this position was first established by the philosophers of the Enlightenment, it has remained pretty much unshaken ever since; for it is hard to see how science, however it may develop, could address such 'ultimate' questions. So although modern scientists may often ask various kinds of 'why?' questions about particular structures or events, the ultimate and most general principles taken to underlie all phenomena are not regarded as admitting of the question 'why is it so?' If we were to achieve a complete and unified theory of the universe (fulfilling the grand philosophical-cum-scientific vision that links Descartes and Hume, Newton and Einstein, right down to present-day cosmologists such as Hawking), such a theory would subsume all observable phenomena in the universe under the fewest and most comprehensive laws or principles; but as to why these principles obtain, this would have to remain, in Hume's graphic phrase, 'totally shut up from human curiosity and inquiry'.[12]

So we have a problem about the modern hope for a grand comprehensive physics that would be the 'ultimate triumph of human reason'. It is an inspiring aim, but one which leaves it very unclear why it should be supposed that a super-theory linking gravity and quantum physics might enable us to tackle the ultimate question of 'why it is that we and the universe exist'. It is sometimes suggested that such a unified theory might turn out to be the only possible theory, in view of the severe constraints that must govern any model that is to be consistent and capable accounting for the universe as we find it. But even if there were to be only one such candidate, it would still be merely the only possible theory given that the universe is as it is – which would still fall short of explaining why there should be a universe at all. Some cosmologists (including Hawking) have speculated that the grand unified theory 'might be so compelling that it brings about its own existence';[13] but it is hard to take this seriously. A theory cannot generate a universe.

SOMETHING RATHER THAN NOTHING

The position we have reached is that while science aims to provide as complete and comprehensive a description as it can of the universe, no matter how successful and unified the theory it ends up with, it cannot explain why there should be a universe there to be explained. We collide with the ancient philosophical question 'Why is there something rather than nothing?' and it seems clear on reflection that nothing within the observable universe could really answer this. If there is a solution to the 'riddle of life in space and time', it would have to lie outside space and time.[14] Here we run into another blank wall: if any such solution must lie beyond the limits of the temporal and spatial universe, outside the 'phenomenal world', as Immanuel Kant called it, then may it not be beyond the horizon of what is humanly knowable? If there is a transcendent realm of the 'noumenal' – something beyond the phenomena, which explains why we and the universe are here – then there is a risk that there will be nothing whatever we can coherently say about it.

We may have reached the limits of science here, but perhaps we have not necessarily reached the limits of human discourse. There is a rich tradition of religious language, both in our Western culture and elsewhere, that grapples with the task of addressing what cannot be fully captured by even the most complete scientific account of the phenomenal world. One might say that it is the task of religious discourse to strain at the limits of the sayable. Some kinds of theology, to be sure, have aimed at keeping entirely within the boundaries of observable evidence and rational demonstration, invoking God as an explanatory hypothesis to account for certain aspects of reality (such as order, design, motion, and so on), rather in the manner of a scientist looking for the best explanation of the data. That enduring strand of natural theology has appealed to many philosophers over the centuries, though it has suffered serious erosion in modern age from the success of rival non-theistic explanations of the relevant phenomena (in particular the triumph of Darwinism). But alongside this quasi-scientific strand in theology, there is also a vast range of religious language that invokes symbol, metaphor, poetry, narrative, and other elements valued for their supposed revelatory power rather than for their strict demonstrative force; religious discourse is here aimed at addressing what cannot fully be put into words, at least into the words of our rational scientific culture, but which can still somehow be shown, disclosed, made manifest.[15]

Such religious discourse gropes towards something beyond the phenomenal world that may give meaning to the universe, and to our human lives. It may not provide a rational scientific solution to the old puzzle of why there is something rather than nothing, for, as we have seen, this is a question which may lie beyond the limits of systematic knowledge. But its advocates would urge that it none the less assuages the vertigo, the 'nausea', as Jean-Paul Sartre called it, that we feel in confronting the blank mystery of existence. The religious answer − one of several responses to the problem of life's meaning to be examined in the pages that follow − aims to locate our lives in a context that will provide them with significance and value. Instead of our feeling thrown into a arbitrary alien world where nothing ultimately matters, it offers the hope that we can find a home.[16]

A RELIGIOUS QUESTION?

Religion is clearly one way in which humans have found a meaning and purpose to their lives. But is it the only way? Albert Einstein asserted bluntly that 'to know an answer to the question "What is the meaning of human life?" *means* to be religious'.[17] That other giant of the twentieth century, Sigmund Freud, also insisted that 'the idea of life having a purpose stands and falls with the religious system'.[18] Yet of course it by no means automatically follows from this linkage that the religious stance is therefore something to be advocated. Freud himself regarded the solution offered by religion as pandering to something unhealthy and disordered in the human psyche:

> The moment a man questions the meaning and value of life, he is sick . . . By asking this question one is merely admitting to a store of unsatisfied libido to which something else must have happened, a kind of fermentation leading to sadness and depression.[19]

Belief in God, according to Freud's view in *Civilisation and its Discontents*, is based on an infantile response: the terrifying 'feeling of helplessness' in childhood aroused the 'need for protection' – for protection through love – which was provided by the father; and the recognition that this helplessness lasts throughout life made it necessary to cling to the existence of a Father, but this time a more powerful one.[20]

This Freudian diagnosis has been highly influential, and can often be seen as informing the idea, voiced by many contemporary atheists, that God is merely a projection formed in response to our human insecurities. But there are at least two problems with this way of dismissing the religious impulse. First, though the abject helplessness of the infant is an apt image of the fragility of the human plight, that fragility, as Freud's own analysis confirms, is clearly not confined to infancy. Our vulnerability, and that of our loved ones, to death, disease and accident is an inescapable part of the human condition; and this being so, to be appropriately aware of it seems precisely what a normal rational human *ought* to be (even granted that constantly dwelling on it may be a sign of neurosis).[21] In the second place, talk of God as a projection does not in the end advance the debate between theists and atheists very much, since it cannot settle the question of whether the impulse to project our longings outwards to an external source does or does not have an objective counterpart. It is certainly plausible that frail and insecure humans would want to project their need for security onto a protective heavenly Father; but a religious believer can equally maintain that since our true destiny lies in union with our creator, we will naturally feel insecure and restless until we find Him. Indeed, precisely this latter theme turns out to be the refrain of many ancient writers on theistic spirituality: *nata est anima ad percipiendum bonum infinitum, quod Deus est; ideo in eo solo debet quiescere et eo frui* – 'the soul is born to perceive the infinite good that is God, and accordingly it must find its rest and contentment in Him alone'.[22] The result of the debate over projection is thus a stand-off: the fact that humans feel a powerful need for God's loving protection logically says nothing either way about whether that protection is a reality.

For the sake of this phase of the argument, however, let us assume for the moment that there is no such divine reality – no objective correlative that could ground our search for life's significance. Would human life, in that case, be empty and pointless? If God is dead, one of Dostoevsky's characters famously declares, everything is permitted;[23] in similar vein, if there is no God, would everything be meaningless?

MEANING AFTER GOD

Depression, so say the experts, is of two kinds, exogenous and endogenous: it can either be triggered by some painful external circumstance, like job loss or bereavement, or it can be apparently spontaneous, presenting as an internal malaise for which there is no immediate outside cause. In a somewhat analogous way, perhaps, it seems that meaning might be either exogenous or endogenous: someone might find their life meaningful in so far as it conformed to the will of a transcendent Creator 'out there' who was the ultimate source of value and significance; but they might instead find meaning 'within', as it were, constructing it from the inside as a function of their own choices and commitments. Friedrich Nietzsche, famous for his announcement of the 'death of God', was clear that humankind, in the post-theistic world, would have to generate significance from within itself – and indeed that this was the only available source for all value: 'Ultimately man finds in things nothing but what he himself has imported into them: the finding we call science, the importing – art, religion, love, pride.'[24]

This conception of meaning as endogenous – the idea of Man as the creator and generator of the meaning of his own life – has plainly had a vast influence on our modern and postmodern culture. The Nietzschean vision can be seen as having three phases. The first is the idea of the 'death of God', which appears in Die Fröhliche Wissenschaft (The Joyful Science, 1882). A madman lights a lantern at midday and runs into the market place crying that he seeks God. He is laughed at by the atheists who are standing around: 'Did he get lost?', they sneer. 'Or has he gone on a journey, or emigrated?' At length the madman announces 'We have killed him, you and I!' And he goes off round the churches of the town to sing a requiem – one that parodies the traditional text of the mass: instead of a prayer for God to grant repose to the dead, it becomes 'requiem eternam Deo' – God himself is consigned to eternal rest.[25]

Over a century later, the shock value of Nietzsche's initial proclamation has faded a little. Walking round the ancient cities of Western Europe, where typically less than 10 per cent of the people now attend religious services, one may feel like echoing the madman's challenge: 'What are all these churches now if they are not the tombs and sepulchres of God?' The culture which once made religious observance so central – in the rites of birth, marriage and death, in the celebration of the changing seasons of sowing and harvest, in the weekly gatherings of the community Sunday after Sunday, in the massive yearly solemnities of Nativity and Resurrection each winter and spring – the culture underpinning all these elaborate structures, if not quite extinct, seems in many places to have either vanished, or be fast crumbling away.

But here the second phase of Nietzsche's predictions comes into play. Just as, after the Buddha was dead, 'his tremendous, gruesome shadow was still shown for centuries in a cave', so, after God is dead, 'there may still be places for thousands of years in which his shadow will be shown – and we still have to vanquish the shadow'.[26] Taking leave of God is not a simple process, like abandoning belief in phlogiston once a better scientific account of combustion comes along. Religious faith does not form an isolated corner of our conceptual map that can be torn off without affecting the main picture; instead (to change the metaphor) it lies at the centre of a vast web of beliefs and attitudes and feelings that are all subtly interconnected. Unravelling them, and coming to terms with the consequences of that unravelling, must involve a radical upheaval, not just in the cognitive sphere, like adjusting or modifying a scientific hypothesis, but in a way that is far more primitive, implying a shift, often at a pre-rational level, in fundamental aspects of our moral, social, aesthetic and psychological orientation towards reality. Large numbers of people may have formally abandoned the idea of God as central to their world-view, but it seems that for relatively few does this feel like having 'arrived'; many instead are left with a sense of vague discomfort, manifested in some by a disquiet about the moral direction of a wholly secular society, in others by intermittent attraction to fashionable alternative modes of spirituality, in others again by a certain melancholic nostalgia for the nourishment and stability of the faith which no longer seems an option. In Nietzsche's eyes, it is as if mankind has acquired a debilitating dependency on the accumulated capital from its religious heritage, and learning to live without the weekly remittance will not be easy. 'Vanquishing the shadow' requires courage and determination.

Here emerges the third phase of the Nietzschean story. For Nietzsche's vision is not a purely destructive one; still less (like the brisk, cheerful atheism purveyed by contemporary secular apologists like Richard Dawkins)[27] is it a plea to sweep away all the religious rubble with the vigorous broom of science that is supposed to clean everything up. Instead, the cry of the madman is imbued with passionate yearning, a fierce lament for the loss of 'the holiest and most powerful of all that the world has yet owned', and a determination to attempt the heroic task of constructing a human surrogate for the defunct God. 'Is not the greatness of this deed too great for us? Must not we ourselves become gods simply to seem worth of it?'[28] Meaning, that vivid sense of purpose without which life slides into flatness and banality, must at all costs be recovered; and to capture this Nietzsche proposes the existential myth of the Eternal Recurrence:

This life as you now live it and have lived it you will have to live once more and innumerable times more; and there will be nothing new in it, but every pain and every joy and every thought and sigh and everything unutterably small or great in your life will have to return to you – and all in the same succession and sequence – even this spider and this moonlight between the trees, and even this moment and I myself. The eternal hourglass of existence is turned over and over, and you with it, a dust grain of dust.[29]

It is not that the envisaged eternal reiteration would somehow bestow objective or external meaning – for what difference could unlimited duration or endless repetition make to the significance of the spider I see in the moonlight? We are indeed alone, in Nietzsche's universe, thrown entirely on our own resources, without any of what he regards as the flabby comforts of religion designed to console the weak. The darkness is all around us, and the only thing that can illumin-ate it is our own indomitable will, a determination to say such a passionate 'Yes!' to each single existential moment of life that even on the condition of eternally repeating it we would chose no other. The question 'Do you want this once more and innumerable times more?' would 'weigh upon your actions as the greatest stress', to be overcome only by an affirmation so powerful that you would 'crave nothing more fervently than this ultimate confirmation and seal'.[30]

Meaning, in Nietzsche's vision, has to be generated entirely from the inside. The world we have to inhabit following the death of God is a world where, in the poet W. B. Yeats' celebrated lines,

> Whatever flames upon the night
> Man's own resinous heart has fed.[31]

MAN, THE MEASURE OF ALL THINGS?

Man, said the philosopher Protagoras, is the measure of all things: of what is, that it is, and of what is not, that it is not.[32] Socrates had little trouble refuting that piece of pretentiousness. Pretentious it is, in its arrogance; the Psalmist's cry 'It is He that hath made us and not we ourselves',[33] whatever one may think of the underlying creed, at least has the humility to acknowledge the basic truth that we exist in the universe as wholly contingent beings, dependent on a reality we did not create. And pretentious too, in its pseudo-profundity. For though Protagoras' modern successors never tire of pointing out that

> there is nothing deep down inside us except what we have put there ourselves, no criterion that we have not created in the course of creating a practice, no standard of rationality that is not an appeal to such a criterion, no rigorous argumentation that is not obedience to our own conventions[34]

the fact remains that none of these human procedures would have any use or value unless they were confronted by an independent non-human reality that in the long run allows those procedures that are effective to flourish, and eradicates those that are faulty. We, mankind, are not the measure of whether a given plant does well in a given soil, or a given engine works more efficiently than another, or the Earth revolves annually around the Sun or vice versa. We create our theories, certainly, but we can only delay, never ultimately prevent, their collapse when they fail to measure up to the bar of actual experience.

Of course there is a residual truth behind the inflated claims of Protagoras and his more sophisticated modern successors. Since we cannot jump outside of our human culture, inspect reality 'as it really is', and then jump back and pronounce such and such a theory true and another false, we should acknowledge that we always have to operate within the context of a continuing dialogue with our peers, with no instant hotline to the truth, no privileged access to a Golden Rule or Procedure that will guarantee that our hypotheses fit reality. Yet to abandon the misguided hope for such guarantees should not lead us to forget that human science aims at discovering (or eliminating) realities that are there (or not there) irrespective of anything we decide. To put it in the more evocative terminology of Yeats' poem, the fuel from our own 'resinous hearts' does not after all 'feed' reality: it illuminates, *but cannot determine*, what is there to be seen upon the night.

How does this bear on Nietzsche's heroic attempt to generate meaning from within? By supposing the unaided human will can create meaning, that it can merely by its own resolute affirmation bypass the search for objectively sourced truth and value, he seems to risk coming close to the Protagorean fallacy. For meaning and worth cannot reside in raw will alone: they have to involve a fit between our decisions and beliefs and what *grounds* those decisions and beliefs. That grounding may, as some religious thinkers maintain, be divinely generated; or it may be based on something else – for example certain fundamental facts about our social or biological nature. But it cannot be created by human fiat alone.

The Nietzschean solution, in short, is untenable; and one may add that it is in any case inhuman, or at least inhumane. For a philosophy that exalts raw will as the key to value and meaning, that makes salvation dependent on the kind of heroic struggle, the greatest stress that can be endured only by the strong, is not likely simultaneously to respect the claims of the tentative, the sceptical, the doubtful, the vacillating, the weak and the helpless – all those who are ill-equipped or disinclined to 'become gods'. Nietzsche, in the kind of rant that sporadically mars his literary and philosophical genius, feverishly welcomed 'the signs that a more manly, a warlike age is about to begin, an age which above all will give honour to valour once again'.[35] Over a century of dire experience later, it would be good to hope that mankind is increasingly seeing reason to prefer the more mundane virtues of compromise and compassion, the less heroic but more democratic values rooted (historically) in the religious ethic of universal brotherhood that Nietzsche scorned. But this postscript on Nietzsche will have to be left hanging for the moment, since it raises general issues about the connection between the quest for meaning and the foundations of morality that will need more time to unravel.

VARIETY, MEANING AND EVALUATION

If human beings cannot create meaning and value merely by an exercise of will, why can't they nevertheless find meaning in the various diverse human activities and projects they undertake? 'Various' and 'diverse' are the key words here. Perhaps the difficulty in the question that baffled the *Hitch Hiker* philosophers Loonquawl and Phouchg was that they were looking for *the* meaning – a single grand all-encompassing answer. That, of course, is the way most religious thinkers have traditionally seen it. But perhaps we need to get away from this spell, and to adopt instead a more modest, piecemeal approach, more in tune with the metaphysically lower-key aspirations of what one might call contemporary secular humanism. Perhaps, as Isaiah Berlin has neatly put it:

The conviction . . . that there exists . . . a discoverable goal, or pattern of goals, the same for all mankind . . . is mistaken; and so too is the notion that is bound up with it, of a single true doctrine carrying salvation to all men everywhere'.[36]

Consider Alan, a golfer. He has retired, has a pension sufficient for his needs, is reasonably healthy, and enjoys above all his thrice-weekly game of golf. Let us assume that he is free from the self-deception and social manipulation that blight the lives of some of his fellow members of the local club: he is not there as a social climber, or to make business contacts, or to show off his expensive golfclubs; he just genuinely enjoys the game. His playing gives his life a structure: each week he looks forward to the coming games, and feels satisfied when they go well. Clearly he has not found 'The Meaning of Life', with capital letters. But why not say, quietly and in lower-case letters, that he has succeeded in finding a meaning, or some meaning, to his life; and that this, and countless similar stories for countless other individuals happily absorbed in their own favoured pursuits, amounts to all that can be said, or needs to be said, on the matter?

Notice that to reach this seemingly modest conclusion certain things need to be assumed. We've stipulated that Alan is comfortably off; and in this designation is included a whole nest of assumptions about a certain easy flow to his life, an absence of too much anxiety and constraint about day-to-day living, an available measure of free time, and an ability to exercise a degree of choice in the use of that time. We have also stipulated that he plays for the sheer joy of the game – uninfluenced by demeaning motives like vain self-importance, or a desire to crawl to the boss; and there is a host of further presuppositions here, this time about the extent to which the chosen pursuit reflects Alan's status as an autonomous agent. If he was playing out of abject fear of losing his job, or because of a subconscious compulsion to surpass his father's sporting achievements, then we would be more doubtful in allowing he had found a meaning to his life – or at least the favourable implications of that phrase would be put in question.

This last point brings out the fact that talk of 'meaning' in life is inescapably evaluative talk. To describe an activity, or a life, as meaningful is evidently to approve or commend it. Now there are many people who have come to think of valuing as a matter of subjective preference; indeed, sentences like 'That's just a value judgement!' are often used to mean something like 'That's no more than your arbitrary personal preference!' But in fact, although there may be some things we just arbitrarily 'take a fancy to', without any objective rhyme or reason, typically we value things *in virtue of objective features* which those things possess. We value a medicine because of its curative properties; we value a piece of music because it is uplifting, or relaxing, or beautifully harmonised; we value a colleague because of her skill or good humour or intelligence. In short, value is typically *grounded* not in arbitrary preference but in objectively assessable features of the world. And characteristically, our value terms reflect this 'grounding' by being what philosophers term 'thick' concepts: they don't just say, thinly, 'wow, that's good!', but rather they carry, packaged-in with them so to speak, those factual features in virtue of which we judge the object to be good.[37] Thus concepts like 'generous' or 'courageous' carry with them a package or checklist of qualifications relevant to the positive evaluation – courage has to do with standing firm in the face of danger, generosity with a certain liberal attitude towards giving to others. Granted for the moment that 'meaningful' is one of these thick concepts – and it certainly seems to be, since to call a life meaningful is not just thinly to say 'wow, it's great!' but to commend it in virtue of some specific features you can point to – let us ask what *is* the package typically associated with this term.

WHAT MEANINGFULNESS IMPLIES

In the first place, to call an activity or a life meaningful normally implies a certain *profundity* or *seriousness* (though 'serious' here need not at all imply 'solemn'). Pursuits can be meaningful in a more or less deep way, but not, as it were, in a shallow way; so to appraise something as meaningful excludes its being trivial or silly. Pastimes like golf appear somewhat borderline here: it seems they can just about qualify as meaningful, but only provided they have a substantial and important recreational function (fostering, perhaps, a certain relaxation, harmony or expansiveness of spirit), or else play some further role, for example by promoting health, or furthering a professional sporting career. But lining up balls of torn up newspaper in neat rows cannot normally be meaningful (except in some special context – maybe it keeps one sane in a prisoner-of-war camp). This links with a second feature: to be meaningful an activity must be *achievement-oriented*, that is, directed towards some goal, or requiring some focus of energy or concentration or rhythm in its execution. Aimlessly throwing darts without any attempt to keep score or any concern for accuracy could not count as a meaningful activity (again, in the absence of special circumstances).

Perhaps the most salient feature of meaningfulness derives from its original semantic home within the domain of language. Meaningfulness is what might be called a hermeneutic concept: for something to be meaningful to a agent, that agent must *interpret* it or *construe* it in a certain way. The words of a marriage ceremony are meaningful because they are construed by the parties as an exchange of promises; exercise in the gym is meaningful because, rather than being an aimless set of bodily movements, it is seen as a programme designed to improve cardiovascular fitness; thrusting a bunch of flowers in someone's hand is meaningful because it is intended as an expression of romantic interest.

Extrapolating from these examples of intentional speech and action, we may draw the further conclusion that meaningfulness in action implies a certain degree of *self-awareness* or *transparency to the agent*: for me to engage in a meaningful activity I must have some grasp of what I am doing, and my interpretation of it must reflect purposes of my own that are more or less transparent to me. This is why someone who is in the grip of psychological distortions or projections, and whose goals are therefore not self-transparent, risks an erosion of their status as an autonomous agent engaged in meaningful activities. Their actions – the obsessive washing of a tablecloth, for instance – may have a deeper resonance that is not properly accessed at the time, so that the agent's own conscious rationalisations of what she is doing ('the household linen must be kept clean') signally fail to justify the endless relaundering of an already spotless cloth. Only when analysis brings to the surface the true significance of the soiled linen (in Freud's famous example, the repressed memory of an embarrassment suffered on the wedding night) is the subject in a position to become self-aware of about her actions, and to regain control of her life.[38] Introducing convoluted cases of this kind is not meant to suggest that all our actions have to be subjected to minute psychoanalytic scrutiny before they meet the transparency condition that allows us to be sure that the way we view our actions does indeed correspond with their true significance. But it is none the less undeniable that the dynamics of human agency are often extremely complex, and that there is therefore something amiss with simplistic accounts that blandly assure us the meaning of a life can be read off as a straightforward function of the goals that an agent consciously declares he has set himself to achieve or the activities he has consciously elected to pursue.

Suppose, however, that an agent is fully engaged, in a self-aware way, undistorted by external manipulation or unconscious projection, on systematic projects that reflect her own rational choice as an autonomous agent. Is this all we need to call their life meaningful? It may be so; but a problem immediately arises that on this showing we could not avoid calling meaningful the life of a dedicated torturer, working devotedly in the service of a corrupt regime. Admittedly, some of those who have reflected on this sort of case have had no difficulty biting the bullet:

> That immoral lives may be meaningful is shown by the countless dedicated Nazi and Communist mass murderers . . . and by people whose rage, resentment, greed, ambition, selfishness and sense of superiority or inferiority give meaning to their lives and lead them to inflict grievous unjustified harm on others. Such people may be successfully engaged in their projects, derive great satisfaction from them, and find their lives . . . very meaningful.[39]

They may perhaps 'find' their lives meaningful, but are such lives *really* meaningful? It would certainly not be a very natural use of our language to bestow the epithet 'meaningful' on the life of the angry, resentful, greedy, ambitious, selfish torturer. But perhaps that is just a cosy parochial prejudice: 'me and my mates' would shrink from calling such a life meaningful, but that might be mere squeamishness, or else an irrational hangover from an outmoded religious world-view – a sentimental wish that only good people can find real meaning in their lives. Can a radically immoral life be really meaningful?

MEANING AND MORALITY

We have already observed that the term 'meaningful' carries with it a package of criteria for its appropriate use (there are problems, for example, in describing trivial projects or the projects of psychologically confused or unself-aware agents, as meaningful). We are now asking whether the life of the dedicated Nazi torturer can indeed qualify as meaningful, and the only way to answer this is to unpack the example more carefully. If such a person is acting out of anger, resentment or a sense of inferiority, for example, it already looks as if they may be falling short of the autonomy and selfhood that is necessary for us to say that their projects represent their own unmanipulated choices about how best to live. A prostitute's life is not meaningful if her 'choice' to go on the streets is the acting out of pain and confusion arising from her being a victim of childhood abuse. The life of the bully who serves the Nazis because he 'derives great satisfaction' from hurting others begins to looks less meaningful when we explore the background and find out that such satisfaction stems from a damaged sense of self-worth produced by his upbringing at the hands of a tyrannical and sadistic father. To be sure, the unhealthy choices and actions described in these sorts of case have a certain sort of significance, just as the stunted growth of a plant sown in polluted ground is significant – it points to something that has gone awry. But it would be wrong to infer, just because the agents find some kind of satisfaction in their activities, that these lives qualify as meaningful in the richer evaluative sense we have identified as belonging to the package typically associated with the label 'meaningful' – the sense that implies an agent's involvement in projects that reflect his or her free and autonomous choices.

But could there not be a completely immoral life that none the less reflected the wholly unmanipulated rational choices of the agent? Could not the rational immoralist – that familiar figure from philosophical discussions of the old question 'Why should one be moral?' – enjoy a perfectly meaningful life? If we make the case one of excessive or per-verted cruelty and inhumanity, then we may be drawn back into questions about the psychological equilibrium of the agent; so let us stipulate instead that the immoralist we have in mind is no monster of vice, nor victim of gross childhood trauma, but is instead just very, very selfish – and selfish, moreover, in a way that signally promotes his chosen projects. He might be like Percy Berkeley, the gourmandising officer in Simon Raven's *Alms for Oblivion* sequence, who 'had not had a thought in his head for twenty years that was not connected with his own immediate pleasure or comfort'.[40] To make the case even harder, let us assume that the chosen projects are very much more impressive than the shallow indulgences of the bon viveur, amounting instead to projects of great cre-ative significance. The much discussed case of Paul Gauguin, who selfishly dumped his family to pursue a self-indulgent but highly creative life in Tahiti, is the case of someone now widely regarded as a painter of genius;[41] and (our critic might ask) if the life of an artistic genius is not meaningful, whose is?

It would be hard to deny that achievements of a high creative order are naturally seen, both by the agents themselves and by onlookers, as meaningful. What holds good for artistic achievement seems to hold good for athletic, technical or intellectual achievement: the great athlete, the brilliant engineer or the gifted mathematician may all feel they have found meaning in their projects and achievements, and may be judged by those around them to lead meaningful lives. Yet in none of these cases does it at first sight appear necessary that the lives in question have to be morally decent lives: just as the artistic genius can be a selfish philanderer, so the great athlete could be a thug, the brilliant engineer a tax dodger, the gifted mathematician a heartless miser, all apparently without affecting the meaningfulness of their lives, so judged. This brings out a disturbing feature of the pluralistic account of life's meaning provided by the brand of secular humanism we are considering. If there is no overarching structure or theory that confers meaning on life, no normative pattern or model to which the meaningful life must conform, then a meaningful life reduces to little more than an engaged life in which the agent is systematically committed to certain projects he makes his own, irrespective of their moral status.

But this is not quite good enough. It is not as if we are speculating in a vacuum about disembodied, deracinated beings who have a clean slate on which to devise the plan that will confer meaning on their lives. We are talking about human lives – the lives of a very special kind of animal, subject to an array of interlinked imperatives – biological imperatives (for food, warmth, shelter, procreation), social imperatives (the need to cooperate, the drive to communicate), emotional imperatives (the need for such things as mutual recognition and affection), and lastly and just as importantly what might be called 'rational imperatives'. Uniquely among known living things we are capable of standing back from our environment, questioning the way things are, challenging the actions of our fellows, entering into dialogues of criticism and justification. In the light of this complex context of interaction and interpersonal dialogue within which we must live our lives, there has to be something unstable about a compartmentalised vision in which individual pursuits and activities can be thought of as bestowing meaning in isolation, irrespective of their moral status, of how they impinge on others. Fulfilment and meaning pursued in ways that involve deceiving or hurting others, or making use of them as mere instrumental fodder for one's own success, closing one's heart and mind to the voice of one's fellow creatures – these are modes of activity that make one *less human*, because the favoured activities have to be conducted at the cost of sealing off one's rational awareness and emotional sensibility so that one is no longer open to such dialogue.

This in turn suggests that to pursue meaning in these inhuman ways risks being self-defeating. Unless the concentration camp guard proposes to turn himself into nothing more than a machine for the infliction of cruelty, he will presumably need, if only in his off-duty hours, human conversation, emotional warmth, the cultivation of friendships, family ties ... Furthermore, since the sensibilities required for such human pursuits cannot be switched on and off at will, but are necessarily a matter of permanently ingrained dispositions of character, the gratification our guard is supposed to be deriving from his gruesome work will inevitably create a psychic dissonance, which will sooner or later to endanger a collapse – either a breakdown of his ability to continue as a torturer or a breakdown of his ability to live a fulfilling home life. Of course it is (unhappily) conceivable that a job that involves cruelty and bullying may produce excitements that may make it horribly attractive to certain individuals; that is not in dispute. The point is that it cannot, for the reasons just given, constitute a coherent model for a meaningful human life.

HUMANITY AND OPENNESS

The factors which constrain our ability as humans to find meaning in our lives depend partly on our emotional make-up and partly on our rational endowments. There is no need to take sides here in an artificial philosopher's battle between David Hume, the champion of sensibility as the basis of morals, and Immanuel Kant, the apostle of rationality. Will anyone who 'wears a human heart', asks Hume, 'tread as willingly on another's gouty toes, whom he has no quarrel with, as on the hard flint and pavement?'[42] There is, on this account, no plausible picture of a truly fulfilled human life cut off from the patterns of feeling that make us naturally disposed to have some minimal concern for our fellow creatures. Can I, asks Kant, rationally conceive of myself as worthy of respect, without recognising as a matter of reason that 'every other rational being conceives his existence on the same rational ground'?[43] Legislating a privilege for oneself which one will not extend to others shows a defective rationality; for to make use of others as a mere means to one's selfish ends is to cut oneself off from the operation of that rational dialogue which defines our humanity. Those determined to take issue with these famous defences of morality could perhaps produce imaginary (and maybe actual) examples of people who insist they have satisfying and meaningful lives despite blocking off their natural sympathies for others, or despite somehow managing to insist on personal privileges they refuse to consider extending to others. But there will always be a certain fragmentation and isolation about such lives. They cannot, by their very nature, flourish into lives that fully embrace our human potentialities for fellow feeling and rational dialogue with others.

If the compartmentalised life is less than fully human, it follows that a truly meaningful life as a human being can be achieved only by one whose pattern of living is in a certain sense *open* rather than closed; that is, whose fundamental dispositions are structured is such a way as not to foreclose the possibility of genuine emotional interaction and genuine critical dialogue with their fellows. This need not mean that in a meaningful life any moral value always has to override all other values: perhaps no one could be a successful artist or scientist or athlete if they were so saintly as to sacrifice all their time and resources to the needs of others. But it does mean that the meaningful life for human beings is an *integrated* life – one where my pet projects and plans are not kept in an isolated category which allows me to pursue them perpetually shielded from the demands on me as a parent, or a friend, or a colleague, or a citizen. Though I may of course have my own special priorities and goals – as I must do, if my life is to be genuinely my own, if I am to be a human agent, not a mere insect in the social hive – [44] nevertheless the walls created by those priorities will never become so thick as to allow me to domineer like a tyrant within the domain of my own creative self-importance.

So what of the Gauguin-type figure? What of the great creator whose string of neglected families and discarded mistresses and betrayed friends are regarded by him as walking wounded, casualties in the all-consuming struggles of a genius? Well, it would be naive and silly to deny that great artists often behave badly. Both Tolstoy and Dickens, though they were in a certain sense 'family men', could in some respects be regarded as husbands from hell. The claim is not that a meaningful life can only be one of untiring virtue: lapses and failures of all kinds are a universal feature of all human lives, as prevalent among the great as they are among the ordinary. Indeed, they may be more prevalent, since the determination required to excel perhaps needs the kind of single-mindedness that has a certain kinship to raw egoism. But one can concede this obvious truth without succumbing to the muddled romantic fantasy that greatness somehow requires or justifies callousness. All the evidence surely points the other way; for great art is great precisely because of its humanity – its heightened vision of the pathos and tragedy and comedy and precariousness of the human condition; and it verges on the absurd to suggest that such a vision is best cultivated through a coarsened and blunted sensitivity to the needs of those fellow humans with whom one is most closely involved. Further reflection along these lines suggests a serious tension, if not downright incompatibility, between the morally insensitive life and the pursuit of artistic creativity. Artistic excellence does not after all operate in a compartment sealed off from the deeper humanity of the artist. That openness we have identified as central to a meaningful life will be as much in point here as anywhere.

This chimes in with an old intuition of Aristotle – that the virtues cannot be fully present in isolation, but are somehow integrated or interconnected.[45] And it accords with the idea to be found in many religious traditions that, in order to be meaningful, life must meet the standards of some pattern tailored to our human nature, rather than being a pure function of isolated individual choice. To the followers of Nietzsche, the champions of creativity and the lordship of the will, such ideas can appear restrictive and confining, like a strait-jacket. But nothing about the idea of the meaningful life as integrated presupposes that every human has to lead the same kind of existence, or that there is not room for many varieties of human flourishing – artistic, athletic, intellectual, and so on. What is presupposed is that to count towards the meaningfulness of a life these varied activities have to be more than just performed by the agent with an eye to personal satisfaction; they have to be capable of being *informed* by a vision of their value in the whole,[46] by a sense of the worthwhile part they play in the growth and flowering of each unique human individual, and of the other human lives with which that story is necessarily interwoven.

The notions here may sound piously high-minded: moral, perhaps even spiritual, values seem to be being invoked as touchstones for a meaningful life. This raises the question: what can be the basis of such exalted conceptions of meaning and value in a universe where, if current scientific orthodoxy is correct, our entire human existence is not much more than a random blip on the face of the cosmos? To that question we now turn.

The Barrier to Meaning

Two

I see these frightening expanses of the universe that shut me in, and I find myself stuck in one corner of this vast emptiness, without knowing why I am placed here rather than elsewhere, or why from out of the whole eternity that has gone before me and the whole eternity that will follow, this one tiny period has been given me in which to live out my life. I see only infinities on every side which shut me in like an atom, like a shadow that lasts only an instant, with no possibility of return. All I know is that I must soon die, but my ignorance is darkest concerning this very death that I cannot avoid.

Blaise Pascal, **Pensées** (1660)[1]

THE VOID

Human beings are hungry for significance. It is intolerable that life should consist merely of one darn thing after another. We want there to be a sense of direction; we would like our lives to constitute an intelligible journey rather than being an aimless drift.[2] If the argument at the end of the last chapter is sound, to be truly meaningful that journey must reflect not just any old purposes or projects we happen to adopt, but those that are genuinely worthwhile. We cannot bestow meaning on our lives just by floundering after individual grat-ification, nor can we create value merely by our own insistent choices, made without regard for the conditions of our (interdependent) flourishing as human beings. A worthwhile life will be one that possesses genuine value – value linked to our human nature and the pursuit of what is objectively conducive to the flowering of that nature. For the theist, the journey that meets these conditions will be the journey of the individual soul towards God; others may construe the journey in less metaphysical terms – as a journey towards enlightenment, or as a quest to realise what is best and noblest in our nature. These ways of characterising the journey all converge on the premise that there are objective values.[3] To put the matter somewhat grandly, a meaningful life will be one oriented as far as possible towards truth and beauty and goodness, or at least by a sense of striving towards those ideals.

But isn't this grand vision hopelessly outmoded? Surely we can only use this high-sounding objectivist language of value if the universe is indeed a moral order,[4] if our strivings are somehow grounded and validated by something 'not ourselves that makes for righteousness'.[5] If 'Truth' 'Beauty' and 'Goodness' have no reality beyond the localised and temporary desires and conventions of humans – those 'imbecile worms of the earth', as Pascal called us[6] – won't it be just a grandiose fantasy to label some lives as more meaningful than any others? True, some lives may appear more or less successful than the average, but from a cosmic perspective it is hard to get very worked up about who loses and who wins, who's in, who's out,[7] when a single fate awaits all. In the words of T. S. Eliot's stark reminder,

O dark dark dark. They all go into the dark,
The vacant interstellar spaces, the vacant into the vacant,
The captains, merchant bankers, eminent men of letters,
The generous patrons of art, the statesmen and the rulers,
Distinguished civil servants, chairmen of many committees,
Industrial lords and petty contractors, all go into the dark,
And dark the Sun and Moon, and the Almanach de Gotha

And the Stock Exchange Gazette, the Directory of Directors,
And cold the sense and lost the motive of action.
And we all go with them, into the silent funeral . . .[8]

There is, in short, a spectre that haunts the seeker for meaning in life: the spectre of the huge, silent, impassive, chillingly
neutral universe that is the backdrop of all our doings. If that
vast blank emptiness, aptly named 'Space', is all the home we
have, then our journey, a journey out of nothing and towards
nothing, risks appearing futile, as void of significance as the
ultimate void that spawned us and will eventually swallow
us up.[9]

Some try to put a brave face on it and argue that we should
not be over-impressed by the vastness that so disturbed Pascal.
Why, they ask, should *size* matter so much? The planet Jupiter
is vastly larger that the Earth, but the most significant object in
the solar system is not that lumbering gas giant, but our own
tiny blue globe – it is the latter not the former that harbours
life and intelligence. Our entire solar system may be a minute
eddy in the huge whirl of the galaxy, and that galaxy merely
one among billions, but so what? Even if we are a tiny blip in
the inconceivably vast expanse of space-time, why should that
make our lives any less meaningful?

Such bravado does not carry conviction. The anguish felt
by Pascal is more than just a fussing about our *smallness* in
comparison to the enormous size of the universe, sobering
though that discovery is. The deeper existential fear has
something to do with the *relation* in which we stand to the vast
whole. In the ancient biblical story of creation, humans have
a secure central role – they are formed in the image of God,
and placed in a world that is in a certain sense brought
into existence for their sakes (thus the Sun and Moon are 'set

in the firmament of heaven to give light upon the earth' (Genesis 1:17)). What Pascal, in the century following Copernicus, was beginning to confront was the stark fact that has dominated cosmology ever since: our existence is not the reason why things are as they are. Indeed, in a certain sense the universe has nothing to do with us. Just as we all have to come to terms with the unpalatable fact that none of us is indispensable – were I to be run over by a bus one day, nothing much would really change – so the 'infinities on every side' would scarcely register the destruction even of this entire planet, along with the 'imbecile worms' who inhabit it.

The vision is doubly bleak when contrasted with what went before. So much of pre-Copernican religious thought was about our relationship to a higher order – vastly greater than us, no doubt, but at least spiritually accessible to us, something towards which we could strive, something with which we could at least in principle aim to be in harmony. This idea is at the root not merely of the Judaeo-Christian world-view, but of many other philosophical systems of antiquity, such as Stoicism: 'In the thought that I am part of the whole', declared the Roman emperor and Stoic philosopher Marcus Aurelius, 'I shall be content with all that comes to pass.'[10] Yet the modern scientific universe glimpsed by Pascal is one which has no relationship at all to our human concerns, our moral and spiritual values, or the direction of our lives. It is just 'out there' – silent, enigmatic. The fear is not about size, but about alienation; shut up, trapped like a speck in a immeasurable cosmos that encloses us but is utterly indifferent to us, each of us can echo Housman's bleak words:

> I a stranger and afraid
> In a world I never made.[11]

THE CHALLENGE OF MODERNITY

Though our lives are indeed utterly fleeting in comparison with the vastness of eternity, perhaps this by itself need not automatically lead to loss of meaning. One might attempt just to bypass such grand cosmic issues and try to find meaning entirely in terms of our ordinary human pursuits and goals (we shall return to this possibility at the start of Chapter Three). But even for those who remain preoccupied with the puny and transient status of humanity vis-à-vis the cosmos, it may be possible to maintain a sense of meaning. Aurelius' approach is a case in point: like all the Stoics he was at pains to stress the ephemeral status of human existence and the inevitability of death, but his underlying belief in the harmonious and rational nature of the universe allowed him to accept mortality with a dignified calmness of mind – and indeed to turn it around to be a positive source of strength:

> Observe how transient and trivial is all mortal life; yesterday a drop of semen, tomorrow a handful of ashes. So spend these fleeting moments on earth as Nature would have you spend them, and then go to your rest with a good grace, as an olive falls in its season, with a blessing for the earth that bore it and a thanksgiving to the tree that gave it life.[12]

Yet that beautiful awareness of life as a gift, and the quasi-religious responses of thanksgiving and blessing, seem very hard for us to recover nowadays. Our sense of ourselves and how we are related to the universe is bleaker and harsher. In understanding this shift, the Copernican revolution that so troubled Pascal's generation is no doubt part of the relevant background; but it is not the whole story. For logically speaking there is no reason why faith in a harmonious and providential universe cannot survive the demoting of the Earth from its unique position. The standard theological doctrine, after all, is that God is an infinite being; so presumably in 'pouring forth his power in creation' he might well be expected to create an unbounded cosmos containing innumerable worlds. And since an infinitely good being must presumably be supposed to have infinite concern for his whole creation – concern extending even to the fall of a single sparrow – then the fact that the Earth may be but one of countless planets supporting life would not be any logical bar to his providential care encompassing all.[13]

What put pressure on traditional theism in the early modern period was less the vastly augmented size of the post-Copernican universe than the sense, fostered by the new mathematical science, of the physical universe as an inexorable machine, obeying its inevitable laws without recourse to human concerns. One of the great seventeenth-century metaphysicians, Benedict Spinoza, concluded that we should abandon altogether the notion of a deity who would intervene in the world for the benefit of mankind, arguing that it is of the essence of 'God or Nature' to preserve a 'fixed and immutable order'.[14] Spinoza's near contemporary, Gottfried Leibniz, concurred in maintaining that the universe is a closed causal system in which all events are 'certain and determined beforehand'.[15] Yet Leibniz went on to argue that the 'sufficient reason' determining every event might reflect *both* an inevitable chain of causes on the physical level *and* the operation of providential purposes on a metaphysical level: there is a 'perfect harmony between the physical kingdom of nature and the moral kingdom of grace, that is to say between God as Architect of the machine of the universe, and God as Monarch of the divine City of Minds'.[16] Though many found this compromise between traditional providentialism and modern science persuasive, there were also fierce critics: in his satirical novel *Candide*, Voltaire scathingly asked how Leibnizian optimism could find a 'sufficient reason' for the crushing of thirty thousand men women and children in the terrible Lisbon earthquake of 1755.[17]

While providentialists could not just blandly shrug off the evidence of such dreadful events, they were not wholly unequipped to respond to the challenge. The existence of suffering in the world had for many centuries been a major theme of religious reflection, and though it evidently shows that we live in a deeply flawed and imperfect world, where all life is transient and vulnerable, this cannot in itself be a logically watertight disproof of the idea of a providential universe. To mention but two traditional lines of defence for the religious apologist (lines reworked in Leibniz's celebrated *Theodicy*, or 'Vindication of God', published in the early eighteenth century), moral evil – wrong and suffering deliberately caused by humans – might be a necessary consequence of God's allowing free action in the world; while natural evil – the suffering caused by physical accident and disease – might be a necessary consequence of unavoidable imperfection in the created order (since a wholly perfect being can only create something other than itself by subtracting from its own perfection).[18] Whatever their merits, these and other strands in the age-old philosophical debate over the problem of evil still continue to form a battleground between theists and atheists, but in the end their impact on the dynamics of religious allegiance turns out to be much less decisive than might be supposed. Those who are able to trust in divine providence, able to commit their lives to a religious interpretation of reality, will be as aware as their atheist counterparts are of the terrible suffering in the world; but they believe they have a way of making sense of that suffering, of finding an under-lying meaning and value in creation, notwithstanding the dreadful ills that arise within it. Those, by contrast, who are unable to make the religious commitment will not see any need to make sense of the destruction and decay all around us, any more than they regard the existence of life itself, or indeed of the universe, as having any ultimate significance: these things simply *there*, and we must cope with the negative aspects as best we can.[19] Leaving these issues unresolved for the moment, let us turn to the impact of later developments in the modern scientific revolution on the traditional religious approach to the meaning of life.

THE SHADOW OF DARWIN

From a historical point of view, the aspects of the rise of modern science so far discussed were enough to make serious waves, but not utterly to overwhelm the traditional concep-tion of a harmonious, providentially ordered universe in which mankind can find meaning. With the Darwinian revo-lution of the nineteenth century there comes an altogether darker and more discordant note, whose repercussions we are still assimilating a century and a half later.[20] Perhaps the most resonant sounding of this sombre new note comes in some musings on evolutionary theory by the great Victorian poet Alfred Tennyson, in a sequence from his masterpiece *In Memo-riam* (a work written in the aftermath of the tragic early death of a dear friend):

> Are God and Nature then at strife
>   That Nature lends such evil dreams?
>   So careful of the type she seems
> So careless of the single life . . .

'So careful of the type?' but no,
      From scarped cliff and quarried stone
      She cries 'A thousand types are gone:
I care for nothing, all shall go.

'Thou makest thine appeal to me:
      I bring to life, I bring to death
      The spirit does but mean the breath
I know no more'. And he, shall he

Man, her last work, who seem'd so fair,
      Such splendid purpose in his eyes
      Who roll'd the psalm to wintry skies,
Who built him fanes of fruitless prayer,

Who trusted God was love indeed
      And love Creation's final law –
      Tho' Nature red in tooth and claw
With ravine, shriek'd against his creed –

Who lov'd, who suffer'd countless ills,
      Who battled for the True, the Just,
      Be blown about the desert dust,
Or seal'd within the iron hills?[21]

The natural world sees countless individuals, and indeed countless species, arising and perishing, with no ultimate purpose, but just the raw brute fact of the continued fight for survival. Man's noble struggle for truth and goodness, the poet fears, may be no more than a temporary blip on the face of a heartless universe, with nothing in the nature of the cosmos to validate that struggle. The belief that Love is 'Creation's final law' is a lost consolation, now untenable. No point in appealing to me, the imaginary voice of nature seems to say, for I am just a brute physical process: 'I bring to life, I bring to death, I know no more'.[22]

'Are God and Nature then at strife?' Philosophically, this might be put as the fear that the evolutionary process disclosed by science is in serious tension with a religious view of the cosmos. But before being swept along to agree that Darwinism threatens to undermine the religious outlook, it is worth pausing to ask exactly why this should be so. It is not as if the idea of the world's gradual evolution burst on the human race halfway through the nineteenth century like a wholly unexpected volcanic eruption. Slow development of the cosmos from a primal state was a possibility that had been canvassed centuries earlier – Descartes, for example, had taken it very seriously in the 1630s, without any discernible erosion of his devout Christian faith;[23] and even if we go back long before the emergence of modernity, we find, for example in Augustine's commentary *On Genesis* (written at the end of the fourth century), ample evidence to suppose that one of the Church's founding fathers would not have found too much difficulty in accepting our current Big Bang cosmology, and even the unravelling of the subsequent story provided by modern evolutionary biology.[24]

It will be helpful at this point to recapitulate the familiar modern evolutionary account of our human nature and origins, in order to see more clearly just what elements of it are supposed to be so threatening to the traditional framework of meaning and value provided by the language of religion. This is roughly how the standard account goes:

> About fourteen billion years ago, the cosmos – all that there
> is, including matter, radiation, space and time –
> somehow began, exploding from a tiny concentration of

matter-space-energy. It expanded very rapidly. Later, gravitational effects caused matter to clump into hydrogen masses (galaxies, stars) and the hydrogen to start fusing into helium. Explosions caused new heavier elements, which formed into planets. On at least one planet, a self-replicating molecule arose. The descendants of this molecule evolved into living organisms, which diversified into all kinds of plants, animals, microbes etc., all solely as a result of natural selection (a totally blind process, operating via random mutation plus a constant struggle for survival). After millions of years one such species became intelligent. Man is a product of these *blind* forces; his moral impulses, like his more savage ones, were shaped by the pressures for species survival. *Religious accounts of our nature and origins have been superseded by the foregoing scientific account, and notions of divine creation are hangovers from primitive, anthropomorphic myths, which attempted to explain our origins in the absence of proper scientific methods.* Science reveals our nature and existence to be the results of entirely natural processes, which we are increasingly managing to understand and predict. We are alone in an impersonal and completely purposeless material universe (though other intelligent beings may have evolved on other planets). One day our solar system will no longer be habitable, and human life (or whatever has evolved from it) will cease, unless it has spread elsewhere. In any case the entire universe will eventually grind to a halt as energy and heat is slowly dispersed through the laws of entropy. *There is no 'ultimate' significance to any of this, nor to any individual human life (indeed, talk of the meaning or purpose of the universe, or of human life, is a philosophical confusion). Abandoning the bogus quest for religious significance, we can nonetheless*

*live better or worse lives depending on how well we manage
to alleviate suffering (through the application of science) and
to maximise the opportunity for enjoyable and enriching
activities for as many people as possible.*[25]

The first thing to underline about the scientific discoveries
listed in this account is that their familiarity should not make
us forget the sheer achievement of modern cosmology in
unravelling the entire physical sequence – Big Bang, expan-
sion, hydrogen, stars, planets, life, intelligence – in terms of a
coherent and consistent pattern of explanation. All of the
diverse events (with the arguable exception of the initial sin-
gularity) are economically and elegantly accounted for in
terms of a very few universal principles of enormous power
and fertility, tested against the bar of systematic observation
and experiment. Since human science is always developing,
future discovery will of course require gaps to be filled, and
revisions, perhaps very substantial ones, to be made; but the
accomplishment to date is still an extraordinarily impressive
one.

Yet what should also be clearly apparent in the above
summary is the very considerable element of interwoven
interpretation (signalled by italics for the relevant sentences)
– interpretation that goes far beyond the truths and hypotheses that pertain strictly to the natural sciences. In particular,
the view of religious thought as something that is *superseded*
or made redundant by the march of science is, whether
you happen to agree with it or not, very evidently a metathesis – a claim that operates at one remove from the claims
of science itself. What reasons are there for subscribing to this
metathesis?

SCIENCE, RELIGION AND MEANING

One possible reason why it might be supposed that science supersedes religion hinges on the conception of the religious idea of divine creation as a *primitive anthropomorphic myth* – one intended to offer an account of how, historically, we got here. God, a kind of superhuman craftsman, made the earth and the heavens and everything in them, rather as a giant carpenter, perhaps, might build a plywood model of the earth and heavens; and God then proceeded to place man there, just as a toymaker might place figurines inside the model, to give it interest. Yet now that we know that everything evolved naturally from a primal explosion, so runs the inference, we can dispense with the cosmic carpenter.

But it only needs to be spelt out in this way for the crudities in such a construal of the language of divine creation to become apparent. Some of the early pagan gods may have been anthropoid beings of the type envisaged here, but the God of Judaeo-Christian-Islamic theism is conceived of very differently – as the source of everything, the source that uttered forth the entire universe out of nothing. An infinite, eternal, creative power that generated the whole physical cosmos would be something of quite terrifying immensity – from a human perspective (to use the term favoured by postmodern theologians), wholly Other. In fact, for any viable version of theism, such a being must not be so Other as to make religious talk entirely incomprehensible: as Thomas Aquinas saw clearly as early as the thirteenth century, theism must at least be able to deploy analogies and models that enable us partially to understand the divine nature, otherwise religious propositions will have nothing whatever to say.[26] But what is clear at the very least is that talk of divine creation should not be (or certainly does not have to be) thought of as an anthropomorphic alternative to the scientific explanations we now have of how the planet Earth and its inhabitants got here.

If, as is reasonable, we take the modern scientific account of the sequence leading to our origins as broadly accurate, then at least one aspect of the account is universally and uncontroversially evident: it presents us with a natural process of staggering power and fertility. Keeping that in mind, consider the kind of ecstatic religious expression used by the eighteenth-century poet Christopher Smart (in lines vividly orchestrated by Benjamin Britten in his cantata *Rejoice in the Lamb*):

> For the TRUMPET of God is a blessed intelligence and so are all the instruments in HEAVEN. For GOD the father Almighty plays upon the HARP of stupendous magnitude and melody . . . Hallelujah from the heart of God, and from the hand of the artist inimitable, and from the echo of the heavenly harp in sweetness magnifical and mighty.[27]

What is the relationship between our scientific account of the cosmos and Smart's awestruck paean of praise? The question is not that easy to answer. Clearly the scientific account does not entail the poetic account, but (and here is the crucial point) it is by no means clear that it is incompatible with it. Thus the theologian Karl Barth, amongst others, has argued that the discourses of science and of religion inhabit different conceptual frameworks, the former dealing with 'the world of men and of time and of things', and the other dealing with the 'world of the Father, and of primal creation and final redemption'.[28] The idea of two realms of discourse, a language of human science and a language of spirituality, is an ancient one, going back as far as the New Testament.[29] Barth's talk of 'different conceptual frameworks' perhaps puts it too strongly if it is taken to imply two hermetically sealed languages with no points of contact whatever; Smart's glorification of the wonders of the cosmos would lose its point if the marvellous phenomena he describes had no relation at all to the universe whose workings are explored by science. But it remains plausible to suppose that the two ways of talking relate to different aspects of a single underlying reality.

Religious language like that of Smart seems (partly at least) a way of *interpreting the significance* of the vast and awesomely complex structure of the physical universe as disclosed by science – a significance expressed not in terms of its physical quantities and mechanical interactions, but as a reflection on its power, its beauty, its rhythm and harmony. This reveals something very important about the terrain occupied by religious language vis-à-vis that occupied by the language of science. It is not as if science says: 'this is how it happened', and religion then offers an *alternative* scenario (for this reason the fundamentalist attempts to get so-called 'Creation theory' taught in American schools as a *rival* to Darwinism seem to involve a radical misconstrual of the relationship between science and religion).[30] The situation, rather, is that the scientist offers an account of how things happened (an account whose acceptability or otherwise depends entirely on tried and trusted criteria such as consistency, coherence, comprehensiveness, predictive power, testing against evidence, and so on); and it then remains a separate (and so far open) question whether the events and processes so established can reasonably be interpreted as manifesting the power and purposes of a divine creator. To use a very crude analogy, a complete physical and chemical analysis of a given set of inkmarks on a page leaves open the semantic question of how the marks are to be interpreted (for example as a sketch or a poem). Interpretations of the latter kind are not a rival to the scientific analysis.

EVOLUTION AND 'BLIND' FORCES

Nothing so far said is supposed to be an argument in favour of a religious interpretation of the existence and evolution of the universe as a manifestation of divine creative action. Equally, none of the scientific truths so far mentioned seem to count decisively against such an interpretation. As a matter of fact, many scientific thinkers and proponents of evolution (from Epicurus onwards) have of course been atheists;[31] but others (from Descartes to the present) have been theists;[32] while others again have taken the agnostic position, maintaining that the findings of evolutionary theory at least do not rule out theistic commitment.[33]

Nevertheless, the specific contribution of Darwin's theory of natural selection (as against the rather vaguer and more schematic gradualism of his predecessors) does seem to raise special difficulties that put pressure on the theist. At the centre of the Darwinian approach, as commonly understood, is the thesis that man owes his origins to a purely accidental chain of essentially blind natural forces. For the crucial point about Darwinian natural selection is that it is not selection at all, in any normal sense. There is, on the Darwinian view, no choice or purposiveness whatever in the natural working of evolution, no picking out of any species or traits, no favouring in the strict sense of that term, but only an entirely impersonal process whereby, inevitably, some will do better than others in the competition for resources, and as a result certain traits will be passed on to the next generation, which would otherwise have been eliminated.

Yet in strict logic there is nothing to prevent such a purely mechanical system (of efficient causality) coexisting with a purposive system (of final causality). The philosopher Leibniz, as we saw earlier, envisaged a realm of spiritual purposes existing 'in harmony' with a realm of physical mechanisms;[34] indeed, he might have gone further and argued that mechanisms and purposes could be co-instantiated in one and the same system – this indeed is what most of us believe about our own brains, which are in a sense blind mechanical systems whose outputs, at the same time, constitute the purposive plannings and doings of conscious agents. There seems no obvious incoherence in such a model being transferred to the evolution of life in the universe as a whole, so that the purely blind mechanical processes leading to organic molecules, and their subsequent development by mutation and natural selection, remains an entirely accurate description of the physico-biological universe, while at the same time being conceived by the theist as instantiating the will of a conscious creator. This is close to the unorthodox solution of Spinoza, for whom the purely physical and mechanical universe (conceived 'under the attribute of extension') exactly corresponds to the meaningful and purposive series of ideas willed by the creator ('under the attribute of thought'), so that we are dealing with 'one and the same thing, but expressed in two ways'.[35]

To be sure, no reason or evidence has so far been advanced to say why we *should* suppose that the physical universe is indeed a manifestation of a divine nature; the point at this stage is merely that the blind and impersonal nature of the evolutionary mechanism, qua purely physical process, does not (as is so often assumed) logically eliminate the possibility of a religious account of its meaning.

A final point is worth making about the sequence of events uncovered by natural science. It is very easy to employ terms like 'accidental' and 'by-product' when referring to the evolutionary account of the origins of life and intelligence – easy, but none the less strictly misleading. For in so far as the great thesis of modern science is precisely that all these complex phenomena (including self-replicating molecules, and the hominid cerebral cortex) arise from entirely natural processes, these should no more be regarded as accidental, or as by-products, than the production of helium from hydrogen in the sun, or the tendency of galaxies to form discs or be grouped in globular clusters. Scientific commentators and popularisers often delight in stressing the sheer contingency of our origins – here we are on a particular planet orbiting a particular star, where the conditions for life just happened to be favourable. Contingent may well be right – but only in so far as any physical event is contingent on the appropriate conditions being present for its occurrence. As to whether we are a cosmic freak, or rarity, it is too early to say. But whether life and intelligence are widely spread throughout the universe or not, one thing is clear: according to the very principles of natural science itself, the universe must be in some sense biophilic and noophilic: the universe is by its nature apt to produce life and intelligence.[36] Our own existence, explained precisely through the standard mechanisms of physics and chemistry, confirms just that.

THE 'NASTINESS' OF THE EVOLUTIONARY MECHANISM

There is a final possible reason why the modern Darwinian story might be thought to undermine theism, though this takes us somewhat beyond the purview of science proper. The thought, nicely encapsulated by Tennyson's image of the *bloodiness* of nature, 'red in tooth and claw', is that the mechanism of competition for survival is simply too grim and ghastly a mechanism to be the manifestation of a supposedly loving creative power, involving too much waste,[37] too much struggle, too much suffering, too much 'munching and crunching' (as one former theist whose disillusionment was due to reflecting on the horrors of natural selection, once put it). This kind of argument takes us away from the crude idea we started with, that theism is superseded by Darwinism, to a far more complex ethical-cum-theological debate.

How we see the natural world will be very much dependent on our basic outlook on reality. 'To the believer', writes one contemporary defender of theism, 'the entire world speaks of God. Great mountains, surging ocean, verdant forests, blue sky and bright sunshine, friends and family, love in its many forms'.[38] To others, the great nineteenth-century atheist philosopher Arthur Schopenhauer for example, the picture is wholly different:

> The futility and fruitlessness of the struggle of the whole phenomenon are more readily grasped in the simple and easily observable life of animals . . . Instead of [any lasting final aim] we see only momentary gratification, fleeting pleasure conditioned by wants, much and long suffering, constant struggle, *bellum omnium* [war among all], everything a hunter and everything hunted, pressure, want, need and anxiety, shrieking and howling; and this goes on *in saecula saeculorum* [world without end], or until once again the crust of the planet breaks.'[39]

Though some may find Schopenhauer's remorseless pessimism exaggerated, there is no denying that the animals, or many of them, do indeed claw and tear at each other in their fight for resources; much of life is indeed a precarious and bitter struggle, and mankind, like the other animals, must make its way against the backdrop of an often hostile environment, where mistakes, in the long or short term, bring severe checks to the flourishing of individuals and whole species. Nature, as Tennyson put it earlier in the poem quoted above, seems 'so careful of the type', yet 'so careless of the single life' – but he goes on despairingly to add that not even *that* is true: she is not even careful of the type, since the archaeological evidence reveals hundreds and thousands of extinct species. Everything seems grist for the remorseless evolutionary mill; nothing is secure. In the face of this, how can we possibly say that 'love [is] creation's final law'?

As far as the violence of much of the animal kingdom is concerned, and the problem it seems to pose for belief in a loving creator, there is perhaps a tendency to exaggerate the savagery involved, and to deploy inappropriate human categories like 'cruel' and 'murderous', when we see a lion tearing at an antelope or a cat chasing and devouring a robin. Many carnivorous species are, to be sure, 'red in tooth and claw', but their behaviour simply follows their hardwired nature, on which their survival depends. Moreover the relatively young science of ecology has taught us that the biosphere is a highly complex and intricately interdependent system: the idea that it would be a nicer or better world where there were only plants and peacefully grazing herbivores may turn out to be shortsighted. It is by no means clear that a benevolent steward of a wildlife reserve would do best to eliminate all the tigers.

The thrust of this ecological response is that a hefty quantum of animal suffering may be an unavoidable ingredient in a flourishing ecosystem, so that it is at least conceivable that a good and loving creator would countenance it. Yet when it comes to human lives, the case (at least according to many peoples' intuitions) seems rather different. We would not respect a ruler who tried to justify the infliction of widespread suffering on his subjects on the grounds that it was a means to some desirable end like ecological stability or variety; the fundamental moral principle of 'respect for persons' could never countenance the use of our fellow humans merely as a means to an end – as mere grist to the ecological mill – in this way.[40] So how, runs the objection, could a supposedly loving and supremely moral creator allow generation after generation of self-conscious and rational beings, made in his own image, to live out their lives in the savagely competitive environment of natural selection, a Darwinian world of 'pressure, want, need, and anxiety'?

Yet to suppose the Darwinian struggle raises a specially acute and new kind of difficulty for the idea of divine authorship of the world seems to involve something of a misconstrual of the way theists have traditionally viewed the created universe. The standard Judaeo-Christian account certainly maintains that the cosmos is good, that every created thing bears a trace of the divine; but a central strand in that account also maintains that our world is a fallen world, that, as St Paul put it, the whole creation 'groans in travail'.[41] The religious view of man is that we are on the move, pilgrims, nomads, on a journey from an imperfect state towards a perfection that is now beyond our reach, but that we can none the less clearly aspire to, and whose outlines we can dimly grasp. 'Here we have no abiding city, but we seek that to come.'[42] It would not be too much to claim that the whole of the religious impulse arises from that profound sense we have of a gap between how we are and how we would wish to be – of something 'deep within our finitude' that points towards the 'infinitude that we crave'.[43] That gap, that fundamental tension or yearning in our nature, does not have to be expressed in the language of theism. But from a theistic perspective, given that the very act of God's creating beings other than himself must involve his creating imperfect beings, it will not be beyond comprehension that the environment he chooses for certain of such beings will be one of stress and struggle, one in which at least some of those creatures will be driven forward on their quest, rather than lapsing into comfort and ease.

MATTER AND SURPLUS SUFFERING

Though stress may have benefits, there is clearly a problem for the theist in that the amount of suffering in the world seems vastly in excess of what could be explained as promoting moral and spiritual growth. Why does not a supposedly omnipotent and benevolent God simply eliminate that surplus? Why not a world with perhaps some stress, but without earthquakes, tidal waves, the anopheles mosquito, anthrax, smallpox, multiple sclerosis . . . ? Part of a possible answer lies in the ancient idea (discussed by Leibniz) of *metaphysical evil*. Even before any question of specific evils or defects, there is, as Leibniz puts it, an 'original imperfection' in the very idea of a created world.[44] This is because it is logically impossible for a perfect being to create something other than itself that is wholly perfect (for a wholly perfect being would just be identical with God). So if he is to create anything at all, God must necessarily create something less perfect than himself; creation necessarily operates, as a long tradition going back to Augustine has it, by what we may think of as a subtraction or diminution from the perfect divine essence.

So far, we have not got very far towards accounting for suffering. For God could presumably create beings that were only slightly less perfect that himself, but would still be blessed immortal, entirely joyful creatures. Indeed, according to many religious traditions he did actually create such beings – the angels. Why not stop there? One response is an idea sometimes dubbed the 'principle of plenitude' – that God's inexhaustible creative power is 'poured forth' in creation.[45] His creativity is inexhaustible: he goes on and on, beyond worlds of light and joy and eternity, to create lesser worlds. At some point down the chain, he creates a material world – the finite world of space and time that we inhabit.

What is a material world? Philosophers, who make frequent use of terms such as 'physicalism' and 'materialism', are often not very good at reflecting on what materiality actually involves. Descartes thought that matter was passive, inert, extended stuff. Locke thought it was solid and impenetrable, rather like lumps of very hard cheddar cheese.[46] But we now know better. Modern science reveals the material world as one of constant fluctuations and interchanges of energy, a shifting shimmering interplay of configurations and forces which are in a constant process of transformation and decay. At the macro level the material universe consists largely of blazing furnaces of hydrogen fusing into helium, their equilibrium delicately poised between the forces of nuclear explosion and gravitational contraction until their fuel is exhausted and they swell up and die. The heavier elements of which planets are composed may clump together in configurations like boulders and mountains which impress humans with their bulkiness and age, but even geologically (let alone cosmically) speaking these are but tiny islands of temporary relative stability in a vast flux of change. The even more fragile biological by-products of all these cosmic processes, microbes and plants and animals, are subject to the same constant change and decay, feeding ultimately off the same fluctuating interplays of energy that characterise the entire cosmos.

Impermanence, instability, decay, in short, are inherent characteristics of material stuff, and everything formed of it – and, crucially, *we humans are formed of this stuff.* The ancient biblical myth says that God formed man 'of the dust of the earth'.[47] Modern science converges on the same idea: living creatures are not some special *sui generis* beings operating according to their own 'vital' principles, but are built of the same sorts of same chemical and physical and microphysical structures as everything else in the natural realm. Despite our intricate complexity, we are an integral part of the material cosmos.

We are now a tad closer to appreciating in a more reflective way the inherent vulnerability of the human condition, the suffering to which we are unavoidably subject. Any creatures inhabiting a material planet, and themselves made of matter, formed of 'the dust of the earth', will necessarily be *mortal*: just like the sun and the stars, and everything else in the cosmos, their life span will be finite, and in an important sense precarious, depending on a delicate balance of fluctuating forces, subject to change and decay, potential prey to instability and collapse.

Could not an omnipotent and wholly benevolent being do something to remedy this? Confronted with what might be called the 'dust of the earth' argument – that our vulnerability is due to our being formed of the inherently vulnerable and unstable elements of material stuff – one might be tempted to voice the curt rejoinder: 'God should have used better dust!' This complaint needs further scrutiny before being accepted as decisive. It seems clear that an omnipotent being would have the power to make beings that are *not* subject to decay and dissolution; as already pointed out, the traditional religious idea of an angelic realm implies that he has in fact done so. But can he make a hydrogen star that does not fuse into helium, thereby exhausting its fuel? One should beware of glib armchair replies like: 'Sure: he could change the laws of atomic physics'. Hydrogen that did not fuse into helium under gravitational compression would not be hydrogen.

More generally, we cannot decide from the armchair whether it is logically possible to create a universe powered (as ours is) by solar energy, but where stars do not decay, where there is no atomic fusion. We may indulge ourselves by fuzzily imagining a world where some of the features science has discovered are held constant while others are changed, but it by no means follows that such a world is a coherent possibility. The fusion of hydrogen into helium is logically linked to a vast web of interconnecting properties of matter: we cannot subtract one without unravelling the whole web. To put the matter in Leibnizian terms: only certain combinations of properties are *compossible* – logically capable of existing in the same universe.[48] God may be omnipotent, but his omnipotence does not extend to making combinations of incompatible properties, like round squares.[49]

The terrible suffering which human beings undergo is sometimes the result of free human acts, and some of it may be morally improving; but much is outside these two categories. It is in our nature as humans to love; yet we lose those we love to death. It is in our nature to desire health and long life; but we and our loved ones fall victim to debilitating diseases. It is in our nature to desire a secure place to dwell; yet the surface of the planet on which we live can often be violent and unstable. And so on. But all these ills that flesh is heir to are inextricably bound up with the fact that we are material beings who live in a material world. We might wish (though very few actually do) that there had been no material world at all; but what we cannot coherently wish is that God had created a material world not subject to change, decay and suffering.

There is a paradox here in that Darwin, so often wheeled in to appear for the prosecution in debates over theodicy, actually now turns up in the schedule of defence witnesses. So long as human life was regarded as a *sui generis* process, specially arranged for our benefit by a cosmic planner, it might have seemed plausible to argue that the illness and pain found in the biosphere were an indictment of the arrangements God had instituted for his creatures. But as soon as human life is seen as continuous with, and part of, the constantly shifting flux of the evolutionary process, with that process itself being a product of the ever-changing cosmic flux of energy exchanges unfolding outwards from the big bang, our human mortality and frailty now appears as manifesting just those features that are inherent to the entire physical cosmos. Suffering, says the ancient Buddhist maxim, is one of the signs of existence – one of the fundamental characteristics of everything that comes to pass in the world. One might put it even more strongly: the very possibility of existence, in any world remotely like ours, depends on mortality, or something closely analogous to it: it depends on the transformation of energy from one form to another, it depends on the undergoing of change and decay, it depends on impermanence and (for that subset of created things that are conscious) it depends on the suffering that is inseparable from that undergoing of change and decay.

There is no space here to embark on the complex issues of theodicy that such a conception raises.[50] The point being made is simply that the dynamic and stressful universe disclosed by modern Darwinism need not be such as to weight the scales overwhelmingly against a theistic interpretation of its significance. Tennyson's revulsion about the ghastliness of the evolutionary struggle is natural enough; but it nevertheless invites reflection on whether to wish for a static and stress-free material universe is a coherent possibility, and whether the crucible of change and impermanence which we in fact inhabit is not logically inseparable from the very idea of a created material universe.

THE CHARACTER OF THE COSMOS

Secundum naturam vivere – to live in accordance with nature – was the Stoic recipe for fulfilment.[51] The idea is essentially a religious one, and is shared by many writers in the Judaeo-Christian tradition. In traditional religious terms, to see our lives as capable of being meaningful is to see them as somehow able to conform to the true nature of the cosmos.

But what is that nature? Many theological writers have maintained that the divine reality is an ultimate reality that utterly transcends the phenomenal world; it is wholly beyond the observable universe investigated by science. But if that is the situation, it is hard to see how we could set about living in harmony with such a reality, or indeed have any relationship to it whatsoever, since it will be entirely beyond our reach and knowledge. In response to this, other theologians have taken a perhaps more promising line involving a compromise, or synthesis, between the idea of God's transcendence and that of his immanence, his indwelling in the cosmos.[52] On this view, although the eternal reality that is God is outside of space and time, his presence is nevertheless also in some manner discernible in the created world: 'O Godhead here untouched, unseen/ All things created bear your trace'.

We do not need to be able to unravel the baffling theology of immanence versus transcendence in order to see a certain intuitive appeal in the lines of the ancient hymn just quoted. If there is an infinite reality behind the universe, it would have to be incomprehensible to finite beings, in the sense that – as Descartes put it – it could not be properly grasped (Latin *comprehendere*) by our finite minds; but perhaps its presence might still be discerned – just as we can glimpse the presence of a distant mountain, even though we are unable, however close we come, to 'grasp' it, to put our arms around it.[53] Indeed, in the standard popular conception of what it is to have a religious outlook on the world, what figures most prominently is precisely this notion of discerning the presence of God in nature:

> a sense sublime
> Of something far more deeply interfused,
> Whose dwelling is the light of setting suns,
> And the round ocean and the living air
> And the blue sky, and in the mind of man.[54]

Yet can that picture survive the rise of modern science? Historians of ideas constantly tell us that the scientific revolution has led to a 'disenchantment' of the natural world. The medieval world-view may have allowed for a world infused with mysterious forces and powers and influences, but the universe as revealed by modern physics is a 'bleached out' universe, an array of interacting particles, void of spiritual qualities of any kind, completely describable in the precise, cold, neutral language of mathematical equations.[55] We don't need the pagan gods any more to set off thunderbolts, to drive the sun across the sky, to enrage and calm the oceans; and we don't even need a super-God, to nudge the whole machine into action.[56] Physical science is entirely autonomous, freed from ghostly thrusters, big and small; and in demystifying it, and accepting its entirely disenchanted nature, human scientific inquiry has finally come of age.

Familiar as this refrain is, it is misleading and arguably fallacious. What is certainly true is that science has led to the elimination of animistic models of the working of nature, replacing them by quantitative and structural descriptions. This is the grand programme for the mathematicisation and mechanisation of science, heralded by Descartes in the seventeenth century and successfully carried through by Newton and his modern successors. If this is the 'disenchantment' of the natural world, then it has been successfully accomplished; but the idea that this has somehow robbed the world of its vitality and beauty, and left us with a dead and colourless universe, a collection of inert mechanical rubble, seems to be a glaring non sequitur. In our inventory of what the universe contains, why should we give special prominence to the rocks and stones? What about what Wordsworth calls the 'living air' – the terrestrial atmosphere, teeming with life? Above all, what about the most prominent inhabitant of that biosphere: how does humankind itself fit into the supposedly bleached out universe of modern science?

The disenchantment view seems infected with a curious dualism, separating off the human observer from nature, as if we form no part of the phenomenon that is under discussion. Yet in so far as we are indubitably part of the cosmos, our own nature must surely have at least some relevance to the question of the nature of that cosmos; and the evidence from the existence of human beings is that the cosmos is such as to produce beings who are eager for truth, receptive to beauty, and who find fulfilment in mutual affection and love. These are very remarkable facts about our universe. They do not of course in themselves establish that our universe as a whole is an ethical order, or that its ultimate source is to be characterised in terms of truth, beauty and goodness. But equally, they give the lie to the glib labelling of our universe as inherently dead and void of value.

As a thought experiment, let us compress the timescale, so that instead of vast stretches of time between the Big Bang and the evolution of intelligent life on earth to its present point, we instead imagine the Big Bang *immediately* giving birth to millions of centres of individual consciousness, each informed with rationality, each yearning for truth, beauty and love. Why does not this manifest the true nature of the cosmos, rather than the standard bleak story about empty space and rubble? The universe is a place of enormous intricacy, beauty and fertility, that much is hard to deny – and certainly nothing discovered by modern science counts against it. Admittedly, we are not in a position to make a direct inference from these facts to the nature of a supposed creator behind the universe: as the great sceptic David Hume correctly pointed out in the eighteenth century, we simply do not have the requisite experience of universes that would enable us to make inferences of this kind to their supposed ultimate cause or basis.[57] But it is not a question here of trying to construct a causal inference to the existence of God; the issue at present is simply that of the discernible character of the cosmos and whether it is at least compatible with a theistic interpretation. Human experience indisputably encompasses features which have traditionally been taken to be signs of the divine presence: the natural world has power to inspire us with its grandeur, with its harmony, with its beauty, and with the warmth and sympathy found in that particular part of nature that is (in Wordsworth's phrase) 'the mind of man'. There are certainly other features that point the other way, such as the competitive viciousness and seeming wastefulness apparent in much of the evolutionary process. So if we are simply looking around us, without any preconceptions either of a theistic or of an atheistic kind, then the observed facts seem to lead to a stand-off when it comes to evaluating the nature of the cosmos we inhabit. An optimistic view of the universe's intrinsic goodness will have some explaining to do, but the same is true of a deeply pessimistic view (like Schopenhauer's) of its inherent vileness. Still the very fact of such a stand-off must leave the door open for the theist (as indeed it does for the atheist, or the agnostic). For it cannot, in any event, be claimed that the universe as disclosed by modern science, and as reflected in our ordinary experience, is *inherently resistant* to a religious interpretation of its significance.[58]

If there is at least the possibility of a religious interpretation of reality, this would open a way for our lives to have meaning in a strong sense that would leave far behind mere local satisfactions of our contingent wants, or the fantasy that we can somehow create our own values.[59] It would provide a model of fulfilment that would locate our human destiny within an enduring moral framework. So far from being a cosmic accident or by-product of blind forces, our lives would be seen as having a purpose – that of attuning ourselves to a creative order that is inherently good. Our deepest responses would be seen as pointing us towards such a goal, and our deepest fulfilment to be attained in realising it.

But of course mere possibility is not enough. *De posse ad esse non valet consequentia*, says the ancient logical maxim: from the fact that something *can* be true, it does not follow that it is *actually* true. Even if there is nothing in the findings of science, or in our ordinary experience, to make the supposition of an ultimate source of value and meaning incoherent, the mere possibility of such a source does not make it rational to act as if it truly existed. To take this further step, we need some way to make the transition from theorising to practice: from theoretical speculation about what might be true, to a reason for living in the hope of its truth.

Meaning, Vulnerability and Hope
Three

The yarns of seamen have a direct simplicity, the whole meaning of
which lies within the shell of a cracked nut. But Marlow was not typical
(if his propensity to spin yarns be excepted), and to him the meaning of
an episode was not inside like a kernel but outside, enveloping the tale
which brought it out only as a glow brings out a haze, in the likeness of
one of these misty halos that sometimes are made visible by the
spectral illumination of moonshine.

Joseph Conrad, **Heart of Darkness** [1902][1]

MORALITY AND ACHIEVEMENT

A meaningful life, it was argued at the end of Chapter One,
must involve worthwhile activities or projects that enable us
to flourish as human beings. Such flourishing requires the
development of our human capacities for feeling and reason:
it involves cultivating the faculties that allow sympathetic
emotional interaction and open rational dialogue with our
fellow humans. This high-minded ideal in turn led us into
questions about the relationship between our moral
endeavours and the nature of the cosmos we inhabit: can the
modern scientific view of the universe leave any room for the
hope that ultimate reality is somehow supportive of our
struggle for meaning and goodness? The conclusion reached
at the end of Chapter Two was that while there is no satisfac-
tory inference from the nature of the world as we find it to the
existence of a supreme underlying principle of meaning and
goodness, nevertheless the character of the world as we find

it cannot be said to *rule out* such an interpretation of its underlying nature.

But do we need to raise these cosmic or religious questions at all? Would God, assuming he exists, really add anything to the validity of those worthwhile projects that develop our human capacities for sympathy and rational dialogue? It is often supposed that God's existence would somehow validate morality or provide it with secure foundations; but in fact (as many philosophers both theistic and atheistic have argued) our moral insights ought to be able to stand alone: they should, if they are worth their salt, be able to command our assent irrespective of whether they are decreed by a Supreme Being. If caring for your neighbour is good, then you ought to be able to *see* its goodness – see that human nature will flower and flourish through such acts of sympathy and concern, see the value of each of us treating others as they would wish to be treated themselves. So if there is a God who commands us to act this way, this must presumably be in virtue of just such features that make the action good; hence God's decrees would not in themselves generate the action's goodness, but simply confirm the goodness we can already recognise there. This is not exactly to say that moral goodness is independent of God – for the theist, nothing is wholly independent of God – but it is to say that, even for the theist, there have to be *reasons* that make things good, and these reasons cannot boil down to the mere fact of their being divinely commanded.[2]

The upshot, not just for the atheist but even for the orthodox theist, is that moral evaluation turns out to have a certain sort of autonomy: the worth and value of our actions is something capable of being discussed and assessed by our human faculties; it cannot be reducible to what is arbitrarily laid down by divine fiat. So why bring religion into the question about the meaning of life at all? It was suggested in Chapter One that a meaningful life is one in which the individual is engaged (without self-deception or other psychological distortions) in genuinely worthwhile activities that reflect his or her rational choice as an autonomous agent. Are not all these elements sufficient to confer meaning on a life, irrespective of the ultimate nature of the cosmos, divine or otherwise?

That seems doubtful. All the elements just mentioned may be *necessary* in order for a life to be meaningful, but something else is also required. In order for a project or activity to give meaning to our lives, we need to feel not just that it reflects our genuine choices and is genuinely worthwhile, but also that it has at least some minimal prospect of success; and conversely, we tend to revise our estimates of meaningfulness if we find that a given project was futile and doomed to failure. Consider David, a millionaire architect, who makes it his life's work to build a hospital in an area where medical facilities are sorely needed. He struggles against great odds to get the project completed, single-mindedly pursing this goal to the point of bankrupting himself, not to mention the neglect of many other rewarding activities that might have engaged his attention. But on the day the hospital is due to be opened, a meteorite hurtles to earth and vaporises the hospital's oil storage tanks; the whole building complex is engulfed in a fireball and razed to the ground, with terrible loss of life. David now bitterly declares that his entire effort was pointless – a tragic and futile waste of energy and resources.

People may try to console him: 'We admire what you tried to do', 'To travel hopefully is better than to arrive', and so

on. But the hard truth is that our assessment of [
project – and this includes the sincere pursu[
worthy goods – is at least partly success-oriente[
it not just to be undertaken in the right spirit, [

something. Moses may never have entered the promised land, but at least he succeeded in freeing his people from slavery, and just before he died, he glimpsed from the hills all the land stretching out before him 'even unto the utmost sea'.[3] His life, he surely felt, had a meaning – it had not after all been wasted. But if he had led his people out merely to die in the desert, never to be heard of again, his dying moments would surely have been very different – beset by thoughts of pointlessness and futility.

FUTILITY AND FRAGILITY

The sense of possible failure and futility that haunts our quest for meaning is of two kinds – universal and particular. From a universal perspective, we know that nothing much in the natural world endures for very long, and that any successes will, at best, only be temporary. The malaise this sense of cosmic impermanence can induce is nicely illustrated in the closing pages of that trusty saga, *The Hitchhiker's Guide to the Galaxy*. The heroes, travelling back in time, find themselves back on Earth among a group of their prehistoric ancestors; they contemplate the grand sweep of human civilisation that lies ahead, but with the vivid knowledge that it will all come to an abrupt end in a few thousand years time, when the planet is brusquely demolished by the Vogon Constructor Fleet to make room for a galactic expressway. With the bene-fit, or rather the curse, of having witnessed the future, they are seized with the thought of how pointless are the eager activities of those early humans all around them, doomed to build a civilisation that will surely perish.[4] Yet of course it does not escape the reader that the future destruction of the Earth by the Vogons is simply a dramatic analogue of the actual fate we know inevitably awaits our planet when the sun expands into a red giant once its hydrogen resources have been consumed. If all human activity is part of a vast inexorable process ending in destruction, then why should anyone make the effort to struggle to achieve what is good and worthwhile, knowing that after a few million turns around the sun, neither right nor wrong nor anything else will survive, as the planet is swallowed up in the inexorable entropic slide of the cosmos towards extinction?

Although this long-term or universal kind of futility is a troubling problem for some reflective people, it can plausibly be argued that a finite stretch of time still provides quite adequate scope for the pursuit of meaningful activities. Virgil did not have to suppose the Roman empire would last for ever in order to proclaim a sense of meaning and pride about its history and prospects. That granted, there nevertheless remains the second possible source of futility, one that operates on a particular rather than a universal level. This is the ever present threat of failure in the lives of each of us as individuals – the plight illustrated by the case of David the architect and his doomed hospital project. What we come up against here is the notorious problem of the frailty of goodness – its seeming lack of robustness in the face of the way the world all too often works. Sages from all faiths and persuasions have long recognised that the path of right action is often beset with obstacles; the struggle to pursue what is good can often just fail, in the face of 'war, dearth, age, agues, tyrannies, despair, law, chance . . .'[5] and all the other ways our hopes can be blighted. Given (as argued a moment ago) that our assessment of the worth of our activities is to some degree success-oriented, it seems that even the most valiant attempt to live a meaningful life will offer many hostages to fortune; indeed, in view of the obstacles which the pursuit of goodness often encounters, it seems that the path to a meaningful life offers an existence fraught with struggle, with chances of achieving a successful outcome that are often decidedly slim – perhaps at best not much more than evens.

None of this, of course, means that a meaningful life will necessarily elude everyone. Sometimes, indeed quite often, projects do work out successfully. So perhaps the conclusion to be drawn is that all human affairs, including our evaluation of people's lives, are subject to an irreducible element of luck.[6] We all know, after all, that 'the best laid schemes o' mice an' men/ gang aft a-gley'.[7] Perhaps we just have to accept that whether the sincere pursuit of worthwhile activities yields a meaningful life will be open to chance: the lucky ones on whom fortune smiles will be able to look back at the end of their lives and pronounce them meaningful, while those who are, by birth, or upbringing, or ill-health, or lack of resources, or accident, unable to pursue worthwhile goals, or prevented from reaching them, will just have to lump it.

While one can hardly fault the consistency of this bleakly restrictive assessment of the percentage of humans able to achieve a meaningful life, the conception it embodies seems both psychologically indigestible and ethically repugnant. It is ethically repugnant because it goes against the long compassionate and egalitarian tradition, rooted in the best of Christian and Islamic thought,[8] that every human creature is eligible for salvation: that the unique dignity and worth of each human being confers infinite value on every one of us, providing us, just in virtue of our membership of the human family, with all we need, provided we turn ourselves sincerely towards the good, to give our lives meaning. And the restrictive conception is indigestible, except perhaps for the most robust of *Übermenschen*, since it expects us, quite unrealistically, to have the confidence to embark on an arduous and demanding voyage with no special reason to hope for a fair wind, no assurance that we have anything beyond our own meagre resources to aid us in the struggle.

'What nonsense', exclaims the valiant surfer gazing out at the towering mountains of water in the bay. 'Even if the ocean will crush me today, or next week, or next year, that does not negate the exaltation I feel as I paddle out to ride the waves.' Human beings have the extraordinary ability to feel joy in the tackling of difficult and challenging tasks. Even Sisyphus, condemned to roll a heavy stone uphill again and again, only to see it each time crashing down to the valley below – even Sisyphus can be thought of as happy as he turns to trudge downhill once more. Albert Camus put it even more strongly: we must think of him as happy – *il faut imaginer Sisyphe heureux.*[9]

But to take the superhuman heroism of the defiant Sisyphus as our model is again the inegalitarian manoeuvre, presupposing the need for a courage so indomitable as to deny realistic prospects for happiness, let alone meaning, to countless numbers of human beings. Most of us, all too conscious of our frailty and vulnerability to fortune, would surely be overwhelmed by the thought that all the cards were as stacked as they were for Sisyphus against the chance of any ultimate success. Yet of course this bleak picture – the 'Absurd' as he calls it – is precisely the picture of the human predicament presented by Camus in *The Myth of Sisyphus*, a book that opens with the chilling pronouncement that 'There is only one really serious philosophical problem, namely suicide'. Life for Camus in this mood could only be absurd, futile and meaningless: lived in a Godless universe, without any of the supporting structures of religion to bolster faith in the power of goodness, it allowed no recourse but 'the refusal to hope and the unyielding evidence of a life without consolation'.[10]

Yet need things be this bleak? Recalling our earlier stress on the value of a life based on the flourishing of our human capacities for sympathy and rational dialogue, could we not see our lives as meaningful, even within a Godless universe, simply in virtue of their being directed towards the *flowering of our human nature*? While the idea of human flourishing seems to me to be vital to any conception of a meaningful life, the problem about dispensing with a religious perspective is that to play its necessary guiding role, human nature in this context has to mean more than just a collection of contingent facts about the sort of creatures we humans have evolved to be: instead, it has to embody a normative ideal of what is noblest and best within us. Yet in that case, appeals to the flourishing of our nature are going to come up against closely similar problems to those already underlined – the problems about the fragility of goodness.

For if we view it from a purely factual standpoint, without any special evaluative focus on the noblest and best, our human nature is simply the result of various configurations of genes produced by a long process of mutation and survival pressure. From this perspective, though we can see how certain moral or altruistic tendencies may have evolved (perhaps because they contributed to cooperation that benefited the species), other traits – aggression, drive for power, ruthlessness – will equally have conferred certain advantages (which no doubt explains the peculiarly vicious and warlike nature of much human history). Now if the ultimate nature of reality contains no bias towards the good as opposed to the vicious, if there is nothing to support the hope that the good will ultimately triumph, if essentially we are on our own, with no particular reason to think that our pursuit of the good is any more than a temporary fragile disposition possessed by a percentage (perhaps a minority) of a certain class of anthropoids – then at the very least it is hard to see how we can achieve the necessary confidence and resolution to follow the path of goodness; and at worst the very idea that some lives can be more meaningful than others begins to seem a fantasy.

The religious perspective – or at least a certain kind of religious perspective (more of this later) – offers the possibility of meaningfulness by providing a powerful *normative framework* or *focus* for the life of virtue. By this I do not mean that religion provides reasons that are needed to justify the goodness of the virtuous life; for (as argued in the previous section) our evaluations of what make actions or lives good are based on features whose worth we can already recognise in human terms, without the need to wait for any divine commands to back them up. The morally good life is indeed one which enables us to fulfil our human nature. But what the religious dimension adds is a framework within which that nature is revealed as more than just a set of characteristics that a certain species happens intermittently to possess, but instead as pointing to the condition that a Being of the utmost benevolence and care that we can conceive of desires us to achieve. Focusing on this dimension, moreover, encourages us with the hope that the pursuit of virtue, difficult and demanding though it often is, contributes however minutely to the establishment of a moral order that the cosmos was created to realise. To act in the light of such an

attitude is to act in the faith that our struggles mean something beyond the local expression of a contingently evolving
genetic lottery; that despite the cruelty and misery in the
world, the struggle for goodness will always enjoy a certain
kind of buoyancy.

RELIGION AND THE BUOYANCY OF THE GOOD

Let us . . . be thankful that our sorrow lives in us as an indestructible
force, only changing its form, as forces do, and passing from pain into
sympathy – the one poor word which includes all our best insight and
our best love . . . For it is at such periods that the sense of our lives
having visible and invisible relations beyond any of which either our
present or prospective self is the centre, grows like a muscle that we
are obliged to lean on and exert.[11]

Religious claims about the buoyancy of goodness are very
easy to misunderstand. Goodness, in the course of actual
human history, is clearly often defeated. When St Paul
encouraged his followers to bear adversity with the cry that
'neither death nor life nor . . . any other creature shall be able
to separate us from the love of God',[12] he cannot have meant
his words to be construed as the naive assertion that things
always work out for the best. The Jewish scriptures, in which
he was so well versed, are packed with stories of terrible trials
suffered by the innocent, of heroic goodness often crushed by
the forces of tyranny and oppression. So the Pauline thought
cannot be a piece of slick optimism, but must involve a more
subtle understanding of the power of Goodness. A rather less
well known passage from his letters perhaps expresses it more
tellingly:

No trial has come upon you that is outside the boundaries of
human experience. And God is faithful, who does not let you

be tested beyond your capacity, but with the trial provides a way out, the power to endure'.[13]

The resilience affirmed here is evidently not a magical overcoming of impossible odds, but a certain mindset which will not judge the value of sticking to the side of goodness by reference to its success or failure measured in terms of outcome, but which generates the courage to endure, irradiated by hope.

'Irradiated by hope.' But on what basis? Are we talking about faith in an afterlife where goodness will be rewarded? Clearly this notion has played a key part in much religious thinking (including much Pauline thought); but without wishing in any way to disparage it, I venture to suggest it does not, in the present context, significantly illuminate our understanding of the religious outlook. Whether or not such an afterlife awaits us is beyond the boundaries of the knowable; but what can be said is that anyone who pursued virtue *solely* in expectation of a personal reward at some future date would thereby automatically have misunderstood the nature of virtue – and indeed would by definition not be acting in a truly virtuous manner. To be religiously motivated to pursue goodness is to strive to act rightly, in a spirit of submission and humility and love, knowing that there is no guarantee of success or comfort, but with an unshakeable conviction that no other path of fulfilment is open to us. Such a mindset is hard to describe in purely cognitive terms; for it is not primarily characterisable in terms of propositions assented to, but is a matter of a certain orientation in which emotions and beliefs and practices of worship and moral convictions merge together in what Wittgenstein called a 'passionate commitment' to a certain form of life.[14]

We have spoken of hope. But the hope involved here is closer to an emotional allegiance to the idea of the power of goodness than to a cognitive attitude of expectation that outcomes will be, on any given occasion, or in general, favourable. Indeed the paradox of religious outlook is that suffering, something which on any rational calculation is most to be avoided, something which we rightly strive to avoid for ourselves and our loved ones throughout our lives, can none the less, in an extraordinary way that defies analysis, function when it does come as the key to a deepening of our nature, bringing us closer to what humans are not yet, but might one day become. This is not to deny that there is much horribly futile and unproductive suffering – comfortable academic theologians who dispute this, coughing dryly into their papery hands, are horrible in their cold glibness. But there is none the less a profound truth grasped in those religions – Christianity is perhaps the most striking example[15] – that put suffering at the very centre of their account of the human condition and the possibility of its redemption.

The resilience of goodness, understood in terms of religious as opposed to secular sensibility, is not a matter of any magical tendency to bounce back or to win through, but rather a matter of something in the human spirit which can respond to the deepest stress and weakness in ways which are transforming. 'The Sacred is revealed to us in the experience of our failure. Religion is indeed the awareness of human insufficiency, it is lived in the admission of weakness.'[16] The old refrain of the Magnificat, *Deposuit superbos et exaltavit humiles* ('He hath put down the proud and exalted the humble'), could crudely be construed as an assertion of a supernatural power that will come along and bash down the arrogantly successful and give the failures a leg up – at least in the next world if not in this. No doubt that is an encouraging and in a sense irreproachable expression of faith for those able to make it their own. But it can also be understood as the insight that through giving up our attachment to the trappings of success, position, money, we become more fully human – more open to the plight of those around us with whom, despite our surface differences, we share so much; such a transformation brings us closer to realising how to live in a world where, sooner or later, we will have to give up everything – our youth, our health, many of those we love, and in the end, even our lives. Status and power temporarily insulate us from our inherent human vulnerability; but in plumbing the depths of that vulnerability we discover what truly matters.

VULNERABILITY AND FINITUDE

Our search for meaningfulness has, perhaps paradoxically, led us to focus on that very precariousness of human life and happiness that one might have supposed to be most inimical to finding meaning in life. Rather than dwelling on such gloomy aspects of the human lot, would it not be better to minimise them, employing all the resources of modern science which has already been so successful in making our existence more comfortable and secure? Yet perhaps the paradox in looking for meaning in a context of precariousness is only apparent. For however far the march of scientific rationality may take us (and obviously it has gained us considerable advances in comfort and physical security), it cannot remove that most fundamental aspect of the human condition – our dependency, our finitude, our mortality. Many philosophers show a strange tendency to conceal this bleak reality from themselves by adopting, almost unconsciously, a kind of jaunty optimism about the powers of human reason. That science can at best mitigate, but can never wholly eradicate, our inherent vulnerability is a fact that they have somehow managed to avoid confronting. This curious self-blinkering extends even to moral philosophers, whom one might have supposed to have a particular responsibility to achieve a clear vision of the conditions of human existence; even more remarkably, it stretches right back to ancient moral philosophers such as Aristotle, that most influential of ethical theorists, who inhabited a world where one might have supposed the fragility of human life to have been all too evident.[17]

The pervasive picture found among moral philosophers, ancient and modern, is of the rational autonomous adult making his self-sufficient decisions about his projects, and loftily deciding the ingredients of the good life. But this is a fantasy. It is a fantasy – not because we are not (partly) rational creatures who are (sometimes) able to take sensible decisions about our welfare, but because it ignores the two poles around which human existence turns, the fundamental facts of our *origins* and of our *destination*. In the first place, the context from which we all emerged, the context in which our goals were mostly shaped and our life's direction largely determined, is a context of abject helplessness and vulnerability – the context of infancy and early childhood. We did not make ourselves, says the Psalmist; we are thrown into the world, says Martin Heidegger: the message is the same – not self-determining autonomy but creaturely dependency. In the second place, the goal and destination of all our elaborate plans and projects is, in the end, nothing. 'O remember how short my time is: wherefore hast thou made all men for nought?'[18] In the long term, as the economist

John Maynard Keynes was fond of observing, we shall all be dead.

Why is the most famous premise in the logic textbooks, that all men are mortal, so important in this context? Part of the answer can be seen to lie in Augustine of Hippo's famous observation that man is 'a kind of intermediate being between the beasts and the angels', *medium quoddam inter pecora et angelos*.[19] Shorn of its metaphysics, what this idea boils down to is the thesis of a perpetual tension in our make-up: we are constrained by our nature, but we see beyond it. Like the animals, we are finite creatures, controlled by a biochemistry that condemns us to inevitable decline and extinction; but in virtue of the faculty of reason we have the ability to distance ourselves in thought from our quotidian existence, and thereby to perceive our finitude in all its starkness. The very ability to see the implications of our finite nature so acutely, means that, alone among the rest of creation, we cannot wholly be at rest, we cannot be entirely at home in the world. The classic human condition of *Angst*, as Heidegger puts it, is one in which one feels *unheimlich* – uncanny, unhomelike. Everyday familiarity collapses. Dasein, the human being, 'enters into the existential mode of the not at home [*Das Nicht-zu-Hause*].'[20] Or as the poet Rilke put it in the 1922 *Duino Elegies* (beating Heidegger to it by five years, and in a much more elegant form):

> *und die findigen Tiere merken es schon*
> *daß wir nicht sehr verläßlich zu Haus sind*
> *in der gedeuteten Welt.*

> and even the animals sense very quickly
> that we are not fully secure, not really at home
> in the interpreted world.[21]

The human condition is paradoxical precisely because it is our nature, qua human beings, to have boundless aspirations which we cannot, qua human beings, fulfil.

Here, it seems to me, is to be found the enduring appeal, despite modern science's many successes, of the idea of spirituality. If the origin of the spiritual impulse is the gap between what we are and what we aspire to be, then since, as long as we remain finite, neither science nor anything else can close that gap, the only available resource will be some kind of radical interior modification which will enable us to come to terms with it. It is just such an interior modification, allowing the possibility of a meaningful life despite our inherent human weakness and mortality, that the great religions have typically aimed to achieve.

SPIRITUALITY AND INNER CHANGE

All spiritual exercises are, fundamentally, a return to the self, in which the self is liberated from the state of alienation into which it has been plunged by [anxiety]. The 'self' liberated in this way is no longer merely our egoistic, passionate individuality: it is our *moral* person, open to universality and objectivity, and participating in universal nature or thought . . . The practice of spiritual exercises implied a complete reversal of received ideas: one is to renounce the false values of wealth, honors, and pleasures, and turn towards the true values of virtue, contemplation, a simple life-style, and the simple happiness of existing . . .

Pierre Hadot, **Philosophy as Way of Life**[22]

The idea of inner change, freeing the self from its alienated state, runs like a clear thread through the great religions of the world, despite vast differences of doctrine and dogma. In the Christian gospels, the call is for *metanoia* – a fundamental shift in outlook, liberating us from anxious care about wealth and position, and leading us to a kind of 'rebirth' in which life will be lived 'more abundantly', as freely as the 'birds of the air' or the 'lilies of the field', unclogged by concern with outward show and image.[23] Much earlier, but in similar vein, the great Jewish prophet Isaiah, had urged 'everyone who is thirsty to come to the waters and buy and eat wine and milk without money and without price' – not to obtain what could be bought for cash, but to get food for the soul: repentance, righteousness, peace.[24] Much later the prophet Mohammed insisted that Allah is not concerned with bodily appearance but 'looks upon your hearts and your deeds', and that 'the most excellent jihad is that for the conquest of the self'.[25] Parallels can be found in ancient Buddhist ethics, with its stress on a path of contemplation designed to free the self from worldly craving and attachment.[26]

Not everyone agrees. Some of the world's influential ethical systems – Confucianism and Aristotelianism are examples – attach considerable importance to maintaining one's proper station in life, with all the concern for wealth and status that this implies. Aristotle's ethical ideal of the *megalopsychos*, the 'great-souled man', is born into a high culture, healthy, intelligent, affluent and calmly confident of his entitlement to honour and esteem.[27] He is certainly virtuous – has carefully cultivated habits of generosity, affection, courage, temperance, and so on – but the elements he relies on to give meaning to his life also place considerable importance on maintaining worldly success. And it follows that *eudaimonia* or fulfilment, at the centre of Aristotle's conception of the good life, is not something that can be confidently assessed until someone's life is over, since an unlucky loss of the wealth necessary to maintain a 'great-souled' lifestyle cannot be ruled out (hence the gloomy Greek maxim 'Call no man happy until he is dead').[28] Apart from its precariousness, the

Aristotelian ideal is also deeply undemocratic (it was no accident that Aristotle condoned institutionalised slavery), requiring a highly hierarchical society to maintain the comforts of the privileged Athenian gentlemen who are taken to be the prime candidates eligible for the good life.

It would be hypocritical for Westerners, or members of what we so casually call the 'First World' to dismiss out of hand Aristotle's tight linkage between virtue and its external material support. It is easy to use the label 'material' in a denigrating way, as if it referred only to mindless obsessions with fashionable clothes or flashy cars; but the fact is that the chosen activities and projects with which most middle-class professionals fill their lives, including many cultural pursuits like travel or opera, not to mention vital back-up facilities like modern healthcare, would all be impossible without a substantial base of material prosperity. Nevertheless, there is something in most of us that is nervously sensitive to challenges about how we are justified in continuing to live our comfortable lives largely blinkered from the hardships of so many of our fellow humans elsewhere on the planet – humans whose labour and resources our own industrial complex often shamelessly exploits.[29] There are of course complex questions of justice, equality and rights involved here – questions on which many gallons of ink have been spilt by moral and political philosophers. The point to be brought out in the present context is this: even if these global problems of fairness and distribution could miraculously be sorted out, even if we were to imagine a perfect economy of global abundance where no one was exploited and material comfort was assured for all, there would still be further questions to be raised about how far a world of utopian prosperity would cater for the spiritual needs of its inhabitants.

This can be called the 'Brave New World problem', after Aldous Huxley's brilliant futuristic fantasy which asked, in effect, whether humans could find meaning in a life of stable economic prosperity where all disease and discomfort were eradicated.[30] One might suppose that alleviating the fragility of human happiness would eliminate any need for the consolations of spirituality. But it doesn't work like this. The development of the deeper sensibilities needed for any real human fulfilment always carries with it an inherent risk. To have a close loving relationship, for example, involves highly complex and fragile intertwinings of affection, trust, conflict, resolution, challenge and change; deep relationships are dynamic, never static – and therein lies the inherent, unavoidable element of risk, the possibility of pain and sadness that dwells 'in the very temple of delight'.[31] Artistic and creative endeavour, intellectual development, genuine adventure, parenthood, and many other areas of life in which we are most fully human, all involve this dynamic character, with the automatically associated possibilities of loss. We could of course dull our sensibilities, drugging our minds with *soma*, Huxley's all-purpose tranquilliser, or filling our time with nothing more challenging than Centrifugal Bumble-Puppy or Electromagnetic Golf; but the clear and persuasive message from *Brave New World* is that such an existence, though *Angst*-free, quickly reduces to the banal and the meaningless.

So what are the goals of spirituality in the actual world, fragile yet wonderful, which we inhabit? We have spoken of coming to terms with the frailty of our human nature and with the precariousness of those things that are most precious to us. All this is crucial enough; but leaving it there perhaps makes the spiritual path sound too much like a mere exercise in compensation, or a mere coping strategy. To be sure, tranquillity of mind, the 'peace that passes all understanding', has always been seen as one of the chief destinations towards which the religious or spiritual road directs us. But if this was just a matter of finding a way to escape from the anguish and risk of human endeavour, then a gram of *soma* ('a gram is better than a damn') might do the job with a lot less bother. An undisturbed mental state, an acceptance, tranquillity, what the Greeks called *ataraxia* ('not-being-churned-up') – these are all part of what spiritual exercises have traditionally been designed to achieve. But such valued states are typically seen as resulting from something else – from a certain kind of *awareness* or *focus* – and it is here that the link with the idea of the meaning of life is at is strongest.

Most forms of spirituality have in common that they aim to turn us away from typical preoccupations such as career, status and the accumulation of wealth, and prepare us instead to focus something extraordinarily simple and yet mysterious: our presence here, at this moment, 'at the still point of the turning world'.[32] Thus do we begin to experience the raw wonder of existence; the sense of life as a blessing and a gift that we can never fully understand.

Such a sense can be achieved without necessarily presupposing an elaborate structure of philosophical or metaphysical theory about the nature of the self or of ultimate reality, as is illustrated by the welling up of simple affirmation felt by a character in one of A. N. Wilson's novels:

> Richeldis did not merely love her husband, family, garden, house. She loved Life Itself. For her, consciousness itself was the deepest mystery. Here, on the first Sunday in October, a miracle was being enacted for her – the sky, the cedars, the lawn – merely because she had two round moist lumps the

size of grapes connected by tenuous sinews to the nervous system. These infinitely fragile things, eyes, which a single blow, a punch, a knife could obliterate, brought before her the whole visual universe and all its beauty. Likewise – sound, smell, taste and touch were, for Richeldis, the causes of increasing wonder, awe and delight. Life brought with it the realisation that nothing could be taken for granted, nothing. This consciousness produced in her a profound inner thankfulness which she would have liked to be able to express. THANKS BE! She longed to cry it out, to sing it. She knew that she was so lucky. But thanks be to whom or to what, she did not know. Perhaps, had there been some little animist shrine in Dunstable or Leighton Buzzard, she would have gone there to pour thankful libations or offer sacrifices of her first fruits. The bullet-hard raspberries in their freezer bag. Such rites would have satisfied in her the sense that she should, in thanksgiving to Great Nature, return something of Itself to Itself. Sometimes she had gone to church with Bartle in this spirit, but the religion of church made absolutely no sense to her.[33]

The passage not only conveys something of the nature of spiritual experience, but also raises the issue of how far it needs to be connected with a religious vehicle for its expression. Richeldis is typical of many modern Westerners for whom the traditional structures of institutionalised religion have ceased to be an emotionally or intellectually acceptable home for their spiritual needs. For George Eliot, writing about a century earlier than Wilson, the traditional vehicle (despite the fact that she was decidedly sceptical about the intellectual basis of Christianity) was still doing its spiritual job:

Adam's thoughts of Hetty did not deafen him to the service;
they rather blended with all the other deep feelings for which
the church service was a channel to him this afternoon, as a
certain consciousness of our entire past and our imagined
future blends itself with all our moments of keen sensibility.
And to Adam the church service was the best channel he
could have found for his mingled regret, yearning and
resignation; its interchange of beseeching cries for help, with
outbursts of faith and praise – its recurrent responses and
the familiar rhythm of its collects, seemed to speak for him as
no other form of worship could have done . . . The secret of
our emotions never lies in the bare object, but in its subtle
relation to our own past: no wonder the secret escapes the
sympathising observer, who might as well put on his
spectacles to discern odours.[34]

Let us pause for a moment to draw the threads together.
Our argument so far has been that the pursuit of meaning
for beings whose existence is inherently fragile requires more
than the rational engagement in worthwhile projects; it
requires a certain sort of religious or quasi-religious mindset.
Involved in this mindset is a turning away from evaluations
based solely on external success, and the cultivation of an
outlook that is affirming of the power of goodness, trusting
and hopeful, and which is focused on the mystery and wonder of existence. This may look like a highly cumbersome
package of attitudes, beliefs and emotions; but it has been the
traditional role of religious systems and practices of spirituality to try to provide a mode of worship capable of being a
vehicle for just that package. Our next step will be to look at
the relationship between the various elements of the package
– in particular how the intellectual or doctrinal components

relate to the actual praxis of worship and meditation. The upshot will be that the doctrinal elements do not in fact have the primacy they are so often assumed to have – a result which goes some way towards removing the sceptical and intellectual obstacles which many people see as blocking a religious solution to the problem of life's meaning.

DOCTRINE AND PRAXIS

He preached a deal about doctrines. But I've seen pretty clear ever since I was a young un, as religion's something else besides doctrines and notions. I look at it as if the doctrines was like finding names for your feelings, so as you can talk of 'em when you've never known 'em, just as a man may talk o' tools when he knows their names, though he's never so much as seen 'em, still less handled 'em.

George Eliot, **Adam Bede**[35]

Talking with people about the fundamental problems of life and its meaning often uncovers a vague nostalgia for religious solutions now felt to be no longer an option. This nostalgia can become acute when life (as it tends to) throws up crises of greater or lesser intensity, or significant landmarks of change – marriage, the birth of children, separation, illness, bereavement. There are also, of course, large numbers of people who want nothing to do with religion – and that is no surprise, since the practitioners of 'religion' are a motley crew which includes in its ranks charlatans, money-grubbing fraudsters and disturbed or manipulative characters hungry for power and influence over others, as well as people of genuine humility and kindness and moral insight. To ask if one is in favour of religion is rather like asking in one is in favour of music – the question is meaningless when the label covers everything from muzak to Mozart. And in the case of religion, whether or not we are drawn to it, and if so in what form, tends to be strongly influenced by our personal experiences, particularly in early life. But among the wide variety of responses, from devotion to repugnance, which the term 'religion' inspires, the attitude of one we may call 'Tim', an educated liberal Westerner, seems to be widespread, perhaps even stereotypical. Tim will admit to those in his confidence that he sometimes wishes he could be religious, and in a sense envies those who have faith. He supposes it must be a comfort in times of trouble; and is sporadically attracted by the dignified and resonant way of dealing with life's rites of passage which is offered by some traditional religious services. Tim is clear, none the less, that it would be intellectually dishonest for him to participate (except as a polite onlooker). The crucial point is that he cannot accept the idea of a God or gods – he rejects the whole concept of supernatural forces that might intervene in human affairs. And the kinds of doctrine one is meant to swallow if one joins a particular faith – for example doctrines about supposed miraculous events like the Incarnation or Resurrection – are ones he could never subscribe to.

On this kind of view, whatever may be the benefits of adopting a spiritual framework for living, that option is closed because of the indigestibility of the doctrines one would be required to accept. But do the doctrines have to loom quite so large? The 'Adam Bede position', expressed in the quotation at the start of this section, is that 'religion is about something else besides doctrines and notions'. In the development of the traditional disciplines of spirituality, both Western and Eastern – the disciplines designed to produce the internal change referred to in our previous section – what is stressed above all for the aspiring devotee is the importance not of doctrines but of *practices*: techniques of meditation and prayer, techniques for self-examination and greater self-awareness, and so on.[36] Rather like the physical routines embraced by the devotee of the gym and the workout, such exercises are aimed at producing a significant change in the subject without necessarily requiring one to absorb an elaborate body of theory.

But do not the practices presuppose *belief*? Surely that is the most central part of a religious outlook? Without in any way disparaging the importance in many peoples' lives of the credal statements they affirm, I want to suggest that belief, in the sense of subscribing to a set of theological propositions, is not in fact central to what it is to be religious. The Pauline writings, to be sure, are full of insistence on *believing* certain things about Jesus, in order to be saved; and many modern preachers take a similar line. But it is, on reflection, quite inconceivable that a good and loving God should make the bestowal of his saving love conditional on whether a given human being was ready to affirm a particular proposition, for example about the inviolability of the Mosaic law, or the precise status of Jesus of Nazareth, or the primacy of the prophet Mohammed. Or that he would exclude from the club of the saved those who conscientiously reject the dogmas specific to any of the three faiths just referred to.

Though the proponents of most religions have over the centuries manifested a worrying tendency to dogmatism and intolerance, the insistence on the centrality of actually standing up and making credal statements has historically been a particular feature of Christianity. Some of this arose from theological disputes that took place several centuries after the death of Christ, though other strands go right back to the influence of Paul on the way the religion developed. Paul had been a devout observer of the Jewish law, a fierce defender of orthodoxy; but as a result of his conversion experience he came to feel that Jesus' message of sacrificial love was one that could not be restricted to a particular nation, but had to be open to all mankind – 'gentile or Jew, bond or free'.[37] As a result, on becoming a Christian, his former fierce defence of Jewish orthodoxy turned into a fierce attack – witness his furious opposition to those who continued to insist circumcision was necessary.[38] But, despite himself, he retained enough allegiance to the mindset of his upbringing to insist that a membership ticket was still required: and that he found in *belief* – the new 'circumcision of the mind' as he called it.[39] So in due course the adoption of a certain belief, in the divinity and resurrection of Jesus, becomes the new circumcision, the new membership ticket; and although, in a powerfully universalist moral vision, the gates are flung open to all nations, nevertheless the enlarged community of the people of God is still an exclusivist club, with those who cannot endure the new mental circumcision cast out.

This is not a treatise on theology, much less Christology, so is not the right place for getting into a complex debate about the status of Jesus – a debate that began during his lifetime when he is reported to have asked his disciples 'Who do you say that I am?'[40] Some historical scholars have argued that as a devout Jew, Jesus of Nazareth could never have claimed to be God (such an idea being an inconceivable blasphemy); and that if (as may have been the case) he thought of himself as a, or the, 'son of God', it is overwhelmingly likely (speaking from a historical perspective) that he did not mean by this that he had no human father, or that he was 'The Son', in the sense of the Second Person of the Trinity.[41] The point to be stressed for the present purpose, however, is that following the path of Christian spirituality can hardly be a matter of sorting out these abstruse Christological debates, or adopting a particular credal stance vis-à-vis the arcane metaphysical formulations of the Council of Nicea (convened several hundred years after Christ's death) concerning the relation between the 'Son' and the 'Father'.[42] What is central to the Christian life is not reaching an intellectual decision on these intricate issues, but rather the adoption of a framework of understanding and praxis. From within that framework, Jesus of Nazareth can appropriately be seen as an 'icon of the invisible God', to use Paul's phrase,[43] someone in whose face the light of the revelation of the glory of God was seen to shine – in his life of self-giving, of love for the helpless and the outcasts, in his work of reconciliation and healing.

The practices of spirituality which stem from this tradition are able to give meaning to the lives of those who adopt them, not in virtue of allegiance to complex theological dogmas but in virtue of a passionate commitment to a certain way of life.[44] What is it to live one's life within this framework? It would be a grotesque arrogance to try to sum this up in any way that claimed to be even remotely definitive, but the following elements at least seem to be discernible parts of the picture.

First, it is to view life as a precious *gift*, not merely 'accidental', but bestowed, stemming from a source that is generative of truth, beauty and goodness.[45] And it is, as a result, to view the world around us (the natural world, and the world of human society) as, despite all its flaws, transfigured by that beauty and goodness and truth.

Second, it is to see one's life as, at certain fundamental turning points, hinging on the choice between good and evil. It is to see individual responsibility as the central fact about who we are. It may be quite compatible with this to feel the pull of the other secular moral frameworks (involving goals such as the maximisation of happiness or the development of human excellences);[46] but the fundamental meaning of our status as moral agents will be seen in terms of a momentous confrontation with our own freedom, our ability to chart the moral course of our lives.

Third, it is to adopt a pattern of life which is structured by traditions of worship – traditions followed not merely in periodic rites of passage (naming, maturity, marriage, death), but also in individual habits of response that mark the daily and weekly rhythms of living: in eating, retiring to sleep, rising in the morning, as well as in collective patterns of song and prayer or meditation, conducted regularly, week by week, and at special seasons and festivals. Such disciplines, or acts of 'submission' (to use an Islamic concept) are adopted not out of a superstitious idea that they will ward off evils or bring good luck, but as a regular focus for moral and spiritual awareness, which itself becomes an engine of interior change (for as Aristotle once put it, in an entirely secular context, by doing certain things one becomes a certain kind of person).[47]

Fourth, it is always to be mindful of the truth affirmed in many central religious texts, that life is made meaningful not by success or material wealth, but by love – and love not of the possessive or appetitive kind, but of the self-giving kind; it is to have the sense, however dim and inadequate, that only in the truly outgoing impulse can a created being transcend itself, and begin to reflect the self-giving radiance of its creator.

FROM PRAXIS TO FAITH

Despite the emphasis so far placed on practical commitments and forms of life, it nevertheless seems very hard to ignore a propositional or cognitive element in the religious outlook. Even a prayer which is doctrinally minimalist, for example the 'Our Father' or 'Lord's Prayer' – a prayer which contains nothing that a devout Jew or Muslim could not perfectly well recite – at the very least implies the proposition that there is a God who stands in a relation to his creatures that is analogous of that of father to children. So whatever the importance of praxis, there seems no getting away from the fact that the practices of spirituality presuppose the truth of certain claims about ultimate reality – about a divine or supernatural source of truth and goodness. This now poses a major problem for those potentially attracted to the spiritual framework as a way of giving meaning to their lives – namely that the truth-claims presupposed may seem ones which cannot responsibly be affirmed. At the start of Chapter One the difficulty was raised that claims about an 'ultimate principle' behind the universe seem by definition to relate to something beyond the phenomenal world, and so outside the scope of human knowledge. How then can we responsibly commit ourselves to a framework which presupposes their truth? The conclusion reached at the end of Chapter Two was that the evidence from the observable world was at best compatible with a claim about its ultimate divine source: although not ruling it out, it was not such as to support it either. Again, in the absence of proper evidential support, how is it responsible to make a belief one's own, or embark on a programme of praxis that presupposes that belief? Must we not agree with the resounding declaration of the nineteenth-century agnostic W. K. Clifford that 'it is wrong always, everywhere, and for anyone, to believe anything on insufficient evidence'.[48].

One philosopher who insisted that statements about God were beyond the reach of human knowledge was Immanuel Kant; but he also famously declared that he 'went beyond knowledge in order to make room for faith';[49] and faith has clearly often been a key element for those adopting the path of spirituality. This fairly obviously applies to the Christian way, but it appears to apply even to non-theistic modes of spirituality such as Buddhism; for as one of its contemporary advocates concedes, with disarming frankness, the Buddhist path is one its adherents are initially drawn to not because its metaphysical claims command their intellectual assent, but rather because they react to them in a more passionate way, first with astonishment, and then with attention and imagination: a Buddhist practitioner does not have to be someone who 'believes in the possibility of *bodhi* [awakening]'; rather, 'the *idea* of *bodhi* can work upon the imagination, can entirely absorb and reorient us, without our believing that there genuinely is such a goal.'[50]

That a religious idea can work on us without its cognitive credentials being first secured is of course an ancient idea. *Praestet fides supplementum*, says the famous hymn of Thomas Aquinas: faith, or trust, supplements the deficiencies of the other faculties.[51] What is particularly relevant to our purposes is the link between the operation of faith and the key strand of spirituality which we have already identified – the primacy which it accords to prescribed practices and techniques such as those of meditation, prayer and self-purification. Blaise Pascal, surely one of those in the philosophical canon who is most aptly described as a 'spiritual' writer, was adamant that this practical dimension must take precedence over the intellectual and the theoretical. Partly anticipating Kant, he declared in a celebrated passage that 'if there is a God, he is infinitely beyond our comprehension, since having neither parts nor limits he bears no relation to us. We are thus incapable of knowing either what his is or if he is.' *Reason, then, cannot decide the matter.* He then went on to argue that the benefits of a religious life were so great that it was rational to undertake practical exercises which would eventually guarantee our sincere adoption of a religious outlook. What was required for this purpose was not further rational argumentation, but rather the training of the emotions:

> Your desired destination is faith, but you do not know the road. You want to cure yourself of unbelief, and you ask for remedies: learn from those who were hampered like you and who now wager all they possess. These are people who know the road you would like to follow; they are cured of the malady for which you seek a cure; so follow them and begin as they did – *by acting as if they believed* [by attending church, and so on]. In the natural course of events this in itself will make you believe, this will train you.[52]

'This will train you': Pascal's original verb is *abêtir*, literally to 'make like the beasts', and this has made it seem to some critics as if Pascal is offering us a degrading prescription for the crushing of critical rationality. The underlying idea is in fact much more subtle: the ancient notion (going back to Aristotle) of the training or habituation of the emotions as part of the path towards a desired goal. We become virtuous adults, says Aristotle, by being trained as children to be virtuous, so that, for example, it eventually becomes natural and automatic for us to feel the right emotions (e.g. courageous or generous emotions) in the appropriate circumstances.[53] You guide young children on the path to the desired destination not, initially, by reasoning with them, since they are not in a position to make the relevant rational evaluations, but by training them, moulding their emotions and conduct until the requisite behaviour becomes second nature.

The fact that emotions are trained, perhaps rather as a singer might train his voice or a tennis player her modes of responding to a serve, does not, however, mean an abandonment of critical rationality, since it remains true that the goal of the training is rationally defensible – and indeed beneficial for all concerned. Pascal's position is that there is a rational (indeed almost utilitarian) argument for religious belief, hinging on the benefits of the religious life. The benefits Pascal actually concentrates on are next-worldly: he develops the famous wager or bet, namely that the sacrifices involved in living a religious life are well worth the chance, however remote, of an infinite reward in the next life. What looks like a crudely self-interested calculation has seemed to many readers harshly out of tune with the very idea of spiritual salvation; but fortunately we do not have go into the ethics of the wager here, since the benefits of the spiritual life that we have been focusing on (see the earlier section on 'Spirituality and Inner Change') are not motivating carrots added on as a later reward, but are instead manifest goods intrinsically linked to the value of the spiritual life as actually lived by its adherents: the care of the soul, tranquillity of mind, release from the false pursuits of egoism and material gain, a closer awareness of the mystery of life, an affirmation of its profundity and its blessings.

The second vital element of the Pascalian strategy, along-side the stress on the benefits of the spiritual path, is the insistence on the passions (or emotions), rather than the intellect, as the key to the religious stance. One is reminded here once again of Wittgenstein's concept of religion, in many respects the heir to Pascal's approach: religious faith is neither rational nor irrational but *pre-rational*,[54] involving not so much intellectual assent to doctrines as a 'passionate commitment' or attitude towards life.[55]

This, then, is the core of the argument I am adapting from Pascal. First, there are clear benefits attached to the spiritual life. Second, the metaphysical doctrines underpinning it relate to matters which are not within the domain of rational knowledge. Third, that is something we need not worry about too much, however, since the adoption of the relevant practices will generate a passionate commitment that bypasses the need for prior rational conviction.

For many philosophers, this seemingly cavalier attitude to the cognitive component of the religious life will be anathema; and what is more, the insistence on the primacy of practice may appear to invert the proper relationship between theory and praxis. Thus one of Wittgenstein's critics has observed that 'it is problematic to say that it is the religious practice, for example in a rite, which gives content to religious doctrines, because the doctrines themselves are supposed to underpin the practice'.[56] I agree that we cannot plausibly suppose that religious practice gives *content* to doctrine; but the idea of the primacy of praxis defended here can be put in different terms. The essential point is that spiritual practices express an existential and a moral response to the human predicament that can plausibly be recognised as beneficial for those who undertake them; and given that such responses can be properly developed only within the context of passionate commitment to the relevant form of life, it makes sense to go the Pascalian route, take the risk, and gradually initiate oneself into the relevant practices, rather than remaining outside in an unsatisfied stance of dispassionate cognitive aloofness.

Let me close this section by taking the example of one, fairly low-key, spiritual practice, the saying of grace before meals. My spokesman is L. R. Kass, author of a late twentieth-century essay on *The Hungry Soul: Eating and the Perfecting of Our Nature*:

> We human beings delight in beauty and order . . . sociability and friendship . . . song and worship. And, as self-conscious beings, we especially crave self-understanding and knowledge of our place in the larger whole . . . The meal taken at table is the cultural form that enables us to respond simultaneously to all the dominant features of our world . . . and the mysterious source of it all . . . Meals eaten before the television set turn eating into feeding. Wolfing down food dishonours both the human effort to prepare it, and the lives of those plants and animals sacrificed on our behalf . . . Especially because modern times hold us hostage to the artificial and the unreal, we do well to remember that the hearth still makes the home, prepared and shared meals still make for genuine family life, and entertaining guests at dinner still nurtures the growth of friendship. A blessing offered over the meal still fosters a fitting attitude toward the world, whose gracious bounty is available to us, and not because we merit it . . . The materialistic view of life, though it may help put bread on the table, cannot help us understand what it *means* to eat . . . Recovering the deeper meaning of eating could help [us] . . . see again that living in a needy body is no disgrace and that our particular upright embodiment orients us toward the beautiful, the good, the true and the holy.[57]

I think it would be difficult for the honest reader not to admit a certain power and resonance in Kass's plea for a more spiritual approach to eating. The secularist might try to suggest that all that is of value here can be reduced to a collection of assorted moral and aesthetic claims about the importance of fellowship and thoughtfulness in eating, and that these can be recognised without spiritual trappings such as saying grace. In reality, as with any spiritual practice that has a symbolic charge (the same goes, in a different way, for any literary or musical or other creative praxis), such reductionism just doesn't work. The relevant performances are not mere trappings, or empty gestures, but (as Kass implies) focusing exercises that provide a vehicle for reflective awareness and emotional response; and if you eliminate the vehicle you risk having no way of reaching the deeper meaning and value of what you are doing. This is true both for everyday rituals like saying grace and for the more momentous rites of passage in life (baptisms, coming of age ceremonies, weddings, funerals): the practices of spirituality generate a resonance, an depth of response, for which there is simply no analogue in the dry language of scientific rationalism or its associated systems of secular ethics.

The Pascalian proposal, that we should embark on the path of spiritual practice, thus offers the hope of finding a dimension of meaning in life which is simply not available within the world-view of analytic rationalism. But despite the stress on praxis and emotional response, as opposed to intellect, there is no ultimate stand-off or conflict here: it is not a matter of wilful dumbing down, or of rejecting the deliverances of rationality. If spiritual practice requires a degree of faith, a willingness to accept something that transcends our cognitive capacities, then such faith can appropriately be described as arising, in Kant's phrase, from a '*need of reason*'.[58] According to Kant, I cannot prove (or disprove) God; yet because it would be humanly impossible to devote my life to the good if I thought I was striving after 'a conception which at bottom was empty and had no object', it is appropriate for 'the righteous man to say "I will that there be a God . . . I firmly abide by this and will not let this faith be taken from me".'

The plea is for an acknowledgement that human beings, in their vulnerability and finitude, need, in order to survive, modes of responding to the world which go beyond what is disclosed in a rational scientific analysis of the relevant phenomena. Such modes of response, moreover, are characteristically expressed through *practices* whose value and resonance cannot be exhausted by a cognitive analysis of propositional contents. Wittgenstein was fond of quoting Goethe's Im *Anfang war die Tat* – 'In the beginning was the Deed'. He also warned, in the closing sentence of his early masterpiece the *Tractatus*, that 'what we cannot speak of we must pass over in silence'. Altering that final phrase, we might say instead that the domain that 'cannot be spoken of' must be *handled through praxis* – the practice of spirituality.[59]

CODA: INTIMATIONS OF MEANING

The Corn was Orient and Immortal Wheat, which never should be reaped nor was ever sown. I thought it had stood from everlasting to everlasting. The Dust and Stones of the Street were as Precious as GOLD . . . And yong Men Glittering and Sparkling Angels, and Maids strange Seraphic Pieces of Light and Beauty! . . . Eternity was Manifest in the Light of the Day and som thing infinit Behind evry thing appeared: which talked with my Expectation and moved my Desire.

Thomas Traherne[60]

Instruction may make men learned, said Bernard of Clairvaux, but feeling makes them wise.[61] Learned academics, whose livelihood is linked to displays of their erudition and cleverness, may act as if all the questions of religion and the meaning of life could be answered from the study or the seminar room. But the illuminations that come from the practice of spirituality cannot be accessed by means of rational argument alone, since the relevant experiences are not available to us during those times when we are adopting the stance of detached rationality.

To *access*, however, is not the same as to *assess*. Even if we cannot gain access to spiritual illumination via rational analysis, this is not to say philosophy, or critical rationality, can have nothing to say about the quest for meaning via the path of spirituality. For philosophy has an obligation to take into account all experience that is part of the human condition. And although claims about the divine may lie beyond the horizon of scientific knowledge, it is not as if embarking on the spiritual quest is a total 'leap in the dark'. For our human awareness, even of the everyday variety, indisputably includes experiences in which spiritual values are made manifest – experiences in which, arguably, we have intimations of a transcendent world of meaning that breaks through into the ordinary world of our five senses. Our apprehension of beauty, the beauty of the natural world, is one example. As simple an experience as that of seeing the *colours* of the leaves in autumn discloses the world around us as resonating with an astonishing harmony and richness; it reveals objects as qualitatively irradiated in modalities which even the most sober of analytic philosophers have agreed are not fully capturable in the language of physics.[62] When William Blake urged us 'To see a World in a Grain of Sand and Heaven in a

Wild Flower',[63] he was not advocating some strange incoherent mindset of the kind philosophers sometimes disparage by using the label 'mystical'. Rather, he was pointing to something that few humans can honestly deny: our ability, in those lucid moments that Wordsworth called 'spots of time',[64] to see the world transfigured with beauty and meaning. There is a clear and unbroken continuum from our immediate everyday experience of the natural world, through the more reflective poetic musings of Blake and Wordsworth, through to the ecstatic vision of Thomas Traherne quoted above, a vision in which the wonder and preciousness of the world, and its human inhabitants, is so vividly manifest.

The pervasive modern vision of the cosmos as bleak and meaningless, with life no more than an accidental scum on the barren rocks, is a vision that is seen through the lens of our own fouled-up lives – the empty concrete wastes of littered parking lots, the dirty, graffiti-defaced walls of decaying warehouses and overcrowded office buildings. But take away the grime our own greed has created, take away the perpetual fog of exhaust fumes and the endless drone of jetliners ripping up the ozone layer, the constant flickering of screens and blaring of speakers. Think back instead only two or three centuries ago, to the limpid scenes captured by Canaletto, or the translucent interiors of Vermeer, the pure air shimmering and sparkling, the colours of everyday objects bright and vivid and new. Think instead of Moses emerging from his tent to gaze up at the brilliant canopy of blazing stars in the clear night of the Sinai desert; the clean pure silence, the astonishing radiance of beauty. *That* is our world: the beauty is not 'projected' onto it by the observer, but is inescapably real, calling forth an irresistible response in our hearts. We *respond* to beauty, as we respond to truth and goodness: as objective realities beyond ourselves, that have the power to inspire us and draw us forward into the light. Of course it is possible, as many philosophers since Hume have argued, that such object-ivity is an illusion, resulting from the mind's tendency to 'spread itself' or project its own feelings outwards onto the world.[65] But it is at least worth considering that talk of 'projec-tion' may apply most aptly *not* to our natural joy and wonder at the immeasurable beauty of the natural world but rather to the bleak modernistic vision of the universe as void of mean-ing and value. It may stem from our own confusions and bitterness as we wilfully turn away from the light, as we stead-ily advance with our bulldozers until we cover the whole planet in concrete and then complain that the cosmos we live in is no more than meaningless rubble.

Our aesthetic experience gives us intimations of a world of value outside our own urgent self-oriented concerns. But to call the wonder expressed in a vision like Traherne's 'aesthetic' is in a certain sense to trivialise it, to make it seem like the precious exclamations of an effete art critic showing off his refinement in a picture gallery. In reality, aesthetic wonder is also suffused with a moral significance: it was no accident that Immanuel Kant linked the brilliancy of the starry heavens above with the moral law within, as the two most awe-inspiring objects in creation.[66] What is manifest in the beauty of the people who walk through Traherne's corn-fields, the young men who are 'angels' and the young women who are 'seraphic pieces of light and beauty', is their human worth: the preciousness of individuals who are vul-nerable, mortal, and yet somehow of eternal value, since their human lot of fragility and suffering has the capacity to deepen their understanding and sympathy – 'that one poor word that involves all our best insight and our best love' –

and ultimately, mysteriously, to provide the grace for redemption and rebirth.

To see the world as Traherne saw it is not something that is dictated by a scientific analysis of the given facts, yet neither is it incompatible with those facts. The moral categories of our experience, so closely bound up with the question of life's meaning, are not arcane or mystical categories, but are inseparable from our human way of being in the world. Inseparable, but not automatically achieved. When things go wrong,

> Life's but a fleeting shadow, a poor player
> that struts and frets his hour upon the stage
> and then is heard no more. It is a tale
> told by an idiot, full of sound and fury,
> signifying . . . nothing.[67]

Macbeth's hell, his deep depression about his life and future, is bound up with a vivid sense of the collapse of any meaning in life. That in turn is triggered by his interior moral collapse, his capitulation to greed and ambition, which lets him take the first step towards betrayal and murder: that capitulation, which was supposed to give him the crown and solve all his problems, turned out to be the first step to ethical disintegration, the first step on the 'primrose path to the everlasting bonfire'. Human beings cannot live wholly and healthily except in responsiveness to objective values of truth and beauty and goodness. If they deny those values, or try to subordinate them to their own selfish ends, they find that meaning slips away.

Perhaps there are some who can achieve a systematic responsiveness to these values without the kind of focus provided by the disciplines of spirituality; but the argument of this book has been that such a 'go it alone' strategy is fraught with problems. We cannot create our own values, and we cannot achieve meaning just by inventing goals of our own; the fulfilment of our nature depends on the systematic cultivation of our human capacities for wonder and delight in the beauty of the world, and the development of our moral sensibilities for compassion, sympathy and rational dialogue with others. Yet, because of the fragility of our human condition, we need more than a rational determination to orient ourselves towards the good. We need to be sustained by a faith in the ultimate resilience of the good; we need to live in the light of hope.[68] Such faith and hope, like the love that inspires both, is not established within the domain of scientifically determinate knowledge, but there is good reason to believe it is available to us through cultivating the disciplines of spirituality. Nothing in life is guaranteed, but if the path we follow is integrally linked, as good spiritual paths are, to right action and self-discovery and respect for others, then we have little to lose; and if the claims of religion are true, then we have everything to gain. For in acting as if life has meaning, we will find, thank God, that it does.

Notes

1 THE QUESTION

1 Douglas Adams, *The Hitchhiker's Guide to the Galaxy* (London: Pan Books, 1979), pp. 135–6.

2 Ludwig Wittgenstein, *Tractatus Logico Philosophicus* [1922] (London: Routledge, 1961), 6.521.

3 Aristotle, *Metaphysics* [c. 330BC], Book I, Ch. 2; cf. Francis Bacon, *The Advancement of Learning* [1605], I, i. 3.

4 For being as an issue, see Martin Heidegger, *Being and Time* [*Sein und Zeit*, 1927] trans. J. Macquarrie and E. Robinson (New York: Harper and Row, 1962), §9; for the 'marvel of marvels', see Heidegger's 'What is Metaphysics' ['Was ist Metaphysic?'], inaugural lecture of 1929.

5 Gottfried Wilhelm Leibniz, *Monadology* [1714], §69.

6 A. E. Housman, 'Tell me not here, it needs not saying', from *Last Poems* [1922], XL.

7 Wittgenstein, *Tractatus*, 6.52.

8 Stephen Hawking, *A Brief History of Time* (London: Bantam Press, 1988), p. 193.

9 Aristotle, *Physics* [c. 325BC], Book II, Ch. 8. The terms 'material', 'formal', 'efficient' and 'final' are not in fact Aristotle's own labels, but are derived from the Latin translations of his works.

10 René Descartes, *Meditations* [*Meditationes de prima philosophia*, 1641], Fifth Replies, in J. Cottingham *et al.* (eds), *The Philosophical Writings of Descartes* (Cambridge: Cambridge University Press, 1985), Vol. II, p. 258.

11 David Hume, *An Enquiry Concerning Human Understanding* [1748], Section IV, part 1; in L. A. Selby-Bigge (ed.) *Hume's Enquiries*, rev. P. H. Nidditch (Oxford: Clarendon, 1975), p. 30.

12 Hume, loc. cit.

13 Hawking, *A Brief History of Time*, pp. 192–3.

14 Cf. Wittgenstein, *Tractatus*, 6.4312.

15 For metaphor as a privileged mode of disclosure, see D. Cooper, *Metaphor* (Oxford: Blackwell, 1986), p. 256.

16 For 'nausea', see Jean-Paul Sartre, *La Nausée* [1936], passim, and *L'Être et le Néant* [1943], trans. as *Being and Nothingness* (London: Methuen, 1957), p. 338. For 'thrownness' (*Geworfenheit*), see Heidegger, *Being and Time*, §29, §38.

17 Albert Einstein, *Mein Weltbild* (Amsterdam: Querido, 1934), trans. S. Bargmann, *Ideas and Opinions by Albert Einstein*, (New York: Crown), p. 11 (emphasis supplied).

18 Sigmund Freud, *Civilisation and its Discontents* [*Das Unbehagen in der Kultur* 1930], Ch. 2; in J. Strachey (ed.), *Standard Edition of the Complete Psychological Works of Sigmund Freud* (London: Hogarth, 1953–74), XXI, p. 76.

19 Letter to Marie Bonaparte of 13 August 1937; in *Letters of Sigmund Freud*, trans. T. and J. Stern (New York: Basic Books, 1960).

20 Freud, *Civilisation and its Discontents*, Ch. 1.

21 Though of course it so far remains an open question whether a religious stance *does* constitute an 'appropriate' way of coming to terms with our vulnerability.

22 St Bonaventure, *Commentarii Sententiarum Petri Lombardi* [1248–55], Book I, 1 iii 2, in *Opera Omnia* (Collegium S. Bonaventurae: Quarachhi, 1891) I, 40. The theme appears earlier, for example in Augustine, *Confessions* [*Confessiones*, 400], Book I.

23 Ivan, in Fyodor Dostoevsky, *The Brothers Karamazov* [1880].

24 Friedrich Nietzsche, *The Will to Power* [*Der Wille zur Macht*, 1888], trans. W. Kaufmann and R. J. Hollingdale (New York: Random House, 1975), p. 327.

25 Friedrich Nietzsche, *The Joyful Science* [*Die Fröhliche Wissenschaft*, 1882], §125; trans. in W. Kaufmann (ed.), *The Portable Nietzsche* (New York: Viking, 1954), pp. 93ff.

26 Nietzsche, *The Joyful Science*, §108.

27 See Richard Dawkins, *River out of Eden: A Darwinian View of Life* (New York: Basic Books, 1995).

28 *The Joyful Science*, §125.

29 *The Joyful Science*, §341. For the existential (rather than literal) construal

of the eternal recurrence, see B. Magnus and K. M. Higgins, 'Nietzsche's works and their themes', in their edited collection *The Cambridge Companion to Nietzsche* (Cambridge: Cambridge University Press, 1996), pp. 37ff.

30 *The Joyful Science*, §341.

31 Everything that man esteems
    Endures a moment or a day.
    Love's pleasure drives his love away,
    The painter's brush consumes his dreams;
    The herald's cry, the soldier's tread
    Exhaust his glory and his might:
    Whatever flames upon the night
    Man's own resinous heart has fed.

(W. B. Yeats, 'Two Songs from a Play', *The Tower* [1928])

The final couplet is connected with the Nietzschean vision by Richard Rorty in the *Times Literary Supplement* no. 5044 (December 3, 1999), p. 11.

32 Plato, *Theaetetus* [c.370BC], 160D.

33 Psalm 100.

34 Richard Rorty, *Consequences of Pragmatism* (Minneapolis: University of Minnesota Press, 1982), p. xlii.

35 *The Joyful Science*, §283.

36 Isaiah Berlin, 'John Stuart Mill and the Ends of Life', in *Four Essays on Liberty* (London: Oxford University Press, 1969).

37 It should be added that even when we use 'thin' concepts ('this object is good', 'that action is right'), our evaluations are still typically grounded in objectively assessable features of the objects or actions in virtue of which they are held to be good or right.

38 Sigmund Freud, *Introductory Lectures on Psychoanalysis* [1916–17], Lecture XVII.

39 John Kekes, *Pluralism in Philosophy: Changing the Subject* (Ithaca: Cornell University Press, 2000), p. 97.

40 Simon Raven, *The Feathers of Death* (London: Anthony Blond, 1959), Ch. 1.

41 For the 'Gauguin problem', see Bernard Williams, *Moral Luck* (Cambridge: Cambridge University Press, 1981), Ch. 2.

42 David Hume, *An Enquiry Concerning the Principles of Morals* [1751], Section V, part 2.

43 Immanuel Kant, *Groundwork of the Metaphysic of Morals* [*Grundlegung zur Metaphysik der Sitten*, 1785], Ch. 2.

44 See John Cottingham, 'The ethics of self-concern', *Ethics* no. 101 (July 1991), pp. 798–817.

45 Aristotle, *Nicomachean Ethics* [c. 330BC], Book VI, Ch. 13.

46 This is not to suggest that meaningfulness always requires explicit reflection on how this integration is achieved, merely that such an integrative story is in principle available.

2 THE BARRIER TO MEANING

1   'Je vois ces effroyables espaces de l'univers qui m'enferment, et je me trouve attaché à un coin de cette vaste étendue, sans que je sache pourquoi je suis plutôt placé en ce lieu qu'en un autre, ni pourquoi ce peu de temps qui m'est donné à vivre m'est assigné à ce point plutôt qu'à un autre de toute l'éternité qui m'a précédé, et de toute celle qui me suit. Je ne vois que des infinités de toutes parts, qui m'enferment comme un atome, et comme une ombre qu ne dure qu'un instant sans retour. Tout ce que je connais est que je dois bientôt mourir, mais ce que j'ignore le plus est cette mort même que je ne saurais éviter.'

(Blaise Pascal, *Pensées* [c. 1660], ed. L. Lafuma
(Paris: Editions du Seuil, 1962), no. 427)

2 It has become fashionable to say that such intelligibility depends on our ability to construct a *narrative* account of our lives; yet since not just any narrative can command our allegiance as providing a meaningful story, the power of narrative seems dependent on prior notions of value and meaning, rather than being itself generative of those notions. See John D. Arras, 'Narrative Ethics', in L. and C. Becker (eds), *Encyclopedia of Ethics* (2nd edn, New York: Routledge, 2001).

3 To avoid misunderstanding, objectivity need not imply a rigid inflexibility. Objectivist accounts of value can perfectly well allow that different cultural, social and economic circumstances may call for different customs and arrangements (for example for differing sexual mores or differing systems for the raising of children). For this point,

see P. Bloomfield, *Moral Reality* (New York: Oxford University Press, 2001).

4 Cf. Simon Blackburn: 'It may once have been a consolation, but it is so no longer, to think that the order of the universe is an ethical order. It is not . . .', *Ruling Passions* (Clarendon: Oxford, 1998), p. 48.

5   The real germ of religious consciousness, therefore, out of which sprang Israel's name for God . . . and which came to be clothed upon, in time, with a mighty growth of poetry and tradition, was a consciousness of the *not ourselves which makes for righteousness*.

(Matthew Arnold, *Literature and Dogma* [1873], Ch. 1, §5)

Though the phrase sounds resolutely objectivist, Arnold in fact argues in this essay against an emphasis on the metaphysical and theoretical content of religious assertions, and instead in favour of the primacy of ethical conduct as the key to the religious impulse.

6   What a chimera then is man. What a novelty. What a monster, what a chaos, what a contradiction, what a prodigy! Judge of all things, imbecile worm of the earth; depository of truth, a sink of uncertainty and error, the pride and refuse of the universe.

(Pascal, *Pensées*, ed. Lafuma, no. 131)

7 William Shakespeare, *King Lear* [1606] Act V, scene 3.
8 T. S. Eliot, from 'East Coker' (1940), III; in his *Four Quartets* [1943].
9   The galaxies will have pretty well evaporated in 10^{18} years, ending the rebirth of stars. After about 10^{40} years there will be no atomic nuclei left, and hence no atoms or molecules . . . The universe will be a very dull place . . . If the Time Traveller journeyed that far forward in time, he would find no beach, no planets, no stars, no atoms, nothing but 'creeping murmur and poring dark'.

(Stephen Weinberg, 'The Future of Science and the Universe', *New York Review of Books*, November 15, 2001, pp. 62–3)

10 Marcus Aurelius, *Meditations* [*Ta eis heauton*, c. AD175], VI 42, trans. M. Staniforth (Harmondsworth, Penguin; 1964).
11 A. E. Housman, *Last Poems* [1922], XII.
12 Aurelius, *Meditations*, IV 48.
13 Such reflections on the status of the Earth in a post-Copernican universe

began as early as Descartes: see his *Conversation with Burman* [1648], ed. J. Cottingham (Oxford: Clarendon, 1976), p. 36.

14 Benedict Spinoza, *Tractatus Theologico-Politicus* [1670], Ch. 6.

15 Gottfried Wilhelm Leibniz, *Theodicy* [1710], I, 52.

16 Leibniz, *Monadology* [1714], §87.

17 Voltaire, *Candide* [1751], Ch. 5.

18 Leibniz, *Theodicy*, I, 20–3.

19 For a judicious selection from the vast literature on the problem of evil, see B.Davies (ed.), *Philosophy of Religion* (Oxford: Oxford University Press, 2000), part V.

20 *On the Origin of Species* has been called 'the most troublesome book published in the second millennium', inaugurating a 'thorough-going secularisation' that, once assimilated, will enable humans to 'give up the last remnants of the idea that they contain a spark of the divine: to see Beethoven and Jefferson as animals with extra neurons'. Richard Rorty, *Times Literary Supplement* no. 5044 (December 3, 1999), p.11.

21 Alfred Lord Tennyson, *In Memoriam* [1850], lvi.

22 Though Charles Darwin's *The Origin of Species* did not appear until nine years after the publication of *In Memoriam*, Tennyson's poem clearly anticipates the idea of a struggle for existence in which countless individuals and species perish. Tennyson took a keen interest in the works of many of Darwin's immediate predecessors, such as Charles Lyell's *Principles of Geology* (1830–3) and Robert Chambers' *Vestiges of Creation* (1844).

23 Descartes, *Le Monde* ['The World' or 'The Universe', 1633], Ch. 6.

24 For Augustine's metaphorical interpretation of creation, see *De Genesi ad litteram* [393], trans. by J. H. Taylor as *The Literal Meaning of Genesis* (New York: Newman, 1982).

25 Though this account is pretty much common ground, much of the phrasing here is drawn from the formulations of others, including especially P. van Inwagen, 'Genesis and Evolution', in E. Stump (ed.), *Reasoned Faith* (Ithaca: Cornell University Press, 1993).

26 Thomas Aquinas, *Summa theologiae* [1266–73], Ia, 13. 5.

27 Christopher Smart, *Jubilate Agno* [c. 1765]; set by Benjamin Britten as *Rejoice in the Lamb*, op. 30 [1943]. The fact that for periods of his life Smart 'teetered on the brink of what his friends regarded as insanity' I. Oudsby (ed.), *The Cambridge Guide to Literature in English* (Cambridge:

Cambridge University Press, 1988), p. 882) should not be taken to detract from the extraordinary incantatory and prophetic power of his writing.

28 Karl Barth, *The Epistle to the Romans* [*Der Römerbrief*, 1919], trans. E. C. Hoskyns (Oxford: Oxford University Press, 1933), p. 29. For a more recent treatment of the idea of scientific and religious language as representing 'two non-overlapping magisteria' or domains of authority, see Stephen Jay Gould, *Rocks of Ages: Science and Religion in the Fullness of Life* (New York: Ballantine, 1999).

29 Compare I Corinthians 2:13; 'We speak not in the words which man's wisdom [*sophia*] teaches, but ... comparing spiritual things with spiritual'.

30 For a review of some of the more influential recent literature in this struggle, see F. Crews, 'Saving Us from Darwin' *New York Review of Books*, October 4 and 18, 2001.

31 The best-known modern example is Richard Dawkins, *The Blind Watchmaker* (London: Longman, 1986), Ch. 1.

32 See Arthur Peacocke, 'Welcoming the "Disguised Friend": Darwin and Divinity', in M. Ruse (ed.), *Philosophy of Biology* (New York: Prometheus, 1998).

33 See M. Ruse, *Can a Darwinian be a Christian?* (Cambridge: Cambridge University Press, 2001), and S. J. Gould, *Rocks of Ages*.

34 Leibniz, *Monadology*, §79. See above, 'The Challenge of Modernity'.

35 Spinoza, *Ethics* [*Ethica ordine geometrico demonstrata*, c.1665], part II, scholium to proposition 7.

36 The term 'biophilic' is used (without any theistic implications) by the astronomer Martin Rees in his book *Our Cosmic Habitat* (London: Weidenfeld & Nicolson, 2002).

37 Why, we must wonder, would the shaper of the universe have frittered away thirteen billion years, turning out quadrillions of useless stars, before getting around to the one thing he really cared about, seeing to it that a minuscule minority of earthling vertebrates are washed clean of sin and guaranteed an eternal place in his company?

(Crews, 'Saving us from Darwin', part I, p. 27)

38 A. Plantinga, 'Religious Belief as Properly Basic', in B. Davies (ed.), *Philosophy of Religion* (Oxford: Oxford University Press, 2000). Extracted

from Plantinga's 'Reason and Belief in God' in A. Plantinga and N. Wolterstorff (eds), *Faith and Rationality* (Notre Dame, IN: University of Notre Dame Press, 1983).

39 *The World and Will and Representation* [*Die Welt als Wille und Vorstellung*, 1818], Book II, Ch. 28; trans. E. F. J. Payne (New York: Dover, 1966), ii, 354. Cf. B. Magee, *Schopenhauer* (Oxford: Clarendon, 1983), Ch. 7.

40 For the principle of respect for persons, see Immanuel Kant, *Groundwork of the Metaphysic of Morals* [*Grundlegung zur Metaphysik der Sitten*, 1785], Ch. 2.

41 'For we know that the whole creation groaneth and travailleth in pain together until now' (Romans 8: 22).

42 'Here we do not have a city (*polis*) that is abiding (*menousa*), but we seek one that is to come (*mellousa*)' (Hebrews 13: 14).

43 A. W. Moore, *Points of View* (Oxford: Clarendon, 1997), p. 278.

44 'There is an *original imperfection in the creature*, before sin, because the creature is limited in its essence' (Leibniz, *Theodicy* [*Essais de théodicée*, 1710], §20).

> Even before the fall, there was an original imperfection or limitation connatural to all creatures ... This, I believe is what the opinion of Augustine and others comes down to, when they maintain that the root of evil lies in nothingness.
>
> (*Discourse on Metaphysics* [*Discours de métaphysique*, 1686], §29)

The term 'metaphysical evil' appears to be due to Leibniz, though (as is clear from his reference to Augustine) the notion goes back to much earlier discussions of the problem of evil. See J. Hick, *Evil and the God of Love* (London: Macmillan, 1966; 2nd edn 1977), p. 13.

45 How do we know that God has not produced an infinite number of kinds of creatures, and thus, as it were, poured forth his power in creation?

> (Descartes, *Conversation with Burman*, p. 36)

Cf. Spinoza, *Ethics*, part I, Appendix and pp. 70ff. The idea is an ancient one, going back to Plato (see *Timaeus* [c. 360BC] 29 E, and Plotinus Enneads [10, 260] v, 2, i and v, 4, i). Cf. J. Hick, *Evil and the God of Love*, p. 21

46 René Descartes, *Principles of Philosophy* [*Principia philosophiae*, 1644], part ii,

art.4. John Locke, *An Essay Concerning Human Understanding* [1689], Book II, Ch.4.

47 Genesis 2: 7.

48 Leibniz, 'A Résumé of Metaphysics' ['Ratio est in natura . . .', c. 1697], in G. H. R.Parkinson (ed.), Leibniz, *Philosophical Writings* (London: Dent, 1973), p.145.

49 On the standard view of God's omnipotence, he is able to do everything that is logically possible (not everything without qualification); cf. Thomas Aquinas: 'Nothing which implies contradiction falls within the omnipotence of God' [*Summa theologiae*, 1266–73], Ia, xxv, 4). Some philosophers however have suggested that even the 'eternal verities' of logic and mathematics depend directly on the will of God; cf. Descartes' letters to Mersenne of 15 April and 27 May 1630, in Cottingham *et al.* (eds), *The Philosophical Writings of Descartes* (Cambridge: Cambridge University Press, 1985), Vol. III, pp. 23, 25.

50 For example, the question of why an omnipotent and benevolent God should not miraculously intervene to prevent the excessive amount of suffering which would otherwise be inevitable in a material world. One possible answer is that his creation of an (unstable, decaying, transitory) material universe necessarily involves a certain loss of control, somewhat analogous to the loss of control involved in the creation of free agents. Another suggested answer is that he chooses not to intervene in order to maintain a certain 'epistemic distance' between himself and his creatures (cf. J. Hick, 'An Irenean Theodicy', in E. Stump and M. J. Murray (eds), *Philosophy of Religion: The Big Questions* (Oxford: Blackwell, 1999)). The first answer may seem to involve some derogation from God's omnipotence; the second from his benevolence.

51 The slogan (a Latin rendering of the Greek *kata physin zen*, the motto of the founder of Stoicism, Zeno of Citium (c. 333–262BC)) is quoted by Cicero in a systematic evaluation of Stoic philosophy in his *On the Ends of Good and Evil* [*De Finibus*, 44BC], Book IV, §26. See A. A. Long and D. N. Sedley, *The Hellenistic Philosophers* (Cambridge: Cambridge University Press, 1987), 63A, 63K.

52 Cf. J. Macquarrie, *Principles of Christian Theology* (London: SCM Press, 1966; 2nd edn 1977).

53 Letter to Mersenne of 27 May 1630; in *The Philosophical Writings of Descartes*, p.25.

54 William Wordsworth, 'Lines composed a few miles above Tintern Abbey', in *Lyrical Ballads* [1798], I, 88.

55 For the 'bleached out' modern universe, see Bernard Williams, *Descartes: The Project of Pure Inquiry* (Harmondsworth: Penguin, 1978), Ch. 10.

56 Pascal, *Pensées*, no. 1001.

57 David Hume, *Dialogues Concerning Natural Religion* [c. 1755; first published posthumously, 1779], part II.

58 Compare W. Rowe (ed.), *God and the Problem of Evil* (Oxford: Blackwell, 2001), part III.

59 Compare the following:

> Our tragedy is that there is no script. Or rather, *we have to write the script ourselves*. We can decide for ourselves which of our inherited values to hold on to, such as loving each other, and which to abandon, like the subordination of women. *And there are new values that we can invent.* Though aware that there is nothing in the universe that suggests any purpose for humanity, one way that we can find a purpose is to study the universe by the methods of science, without consoling ourselves with fairy tales about its future or about our own.
>
> (Stephen Weinberg, 'The Future of Science and the Universe' [emphasis supplied])

3 MEANING, VULNERABILITY AND HOPE

1 Joseph Conrad, *Heart of Darkness* [1902], Ch. 1.

2 For the complex question of the relation between religion and morality (an issue that goes back to Plato's dialogue the *Euthyphro*), see E. Stump and J. Murray (eds), *Philosophy of Religion: The Big Questions* (Oxford: Blackwell, 1999), Part Six.

3 Deuteronomy 34: 2.

4 Douglas Adams, *The Restaurant at the End of the Universe* (London: Pan Books, 1980).

5 John Donne, *Holy Sonnets*, VII, from *Divine Poems* [c. 1620].

6 For some of the substantial literature on this issue, see D. Statman (ed.), *Moral Luck* (Albany: State University of New York Press, 1993).

7 Robert Burns, 'To a Mouse' [1785].

8 Cf. Paul, Epistle to the Colossians [c. AD50], 3:11. For Islamic universalism, cf. The Quran [seventh century CE], 2:136.

9 Albert Camus, *The Myth of Sisyphus* [*Le mythe de Sisyphe*, 1942], tr. J. O'Brien (Harmondsworth: Penguin, 1955, rev. 2000), p. 111.

10 Camus, *The Myth of Sisyphus*, p. 59.

11 George Eliot, *Adam Bede* [1859], Ch. 50.

12 Romans 8:38.

13 I Corinthians 10:13.

14 Cf. Wittgenstein, *Culture and Value* [*Vermischte Bemerkungen*, 1977] (Oxford: Blackwell, 1988) p. 61. See also H. Glock, *A Wittgenstein Dictionary* (Oxford: Blackwell, 1996), s. v. 'religion'. Compare Rudolph Carnap:

> The (pseudo) statements of metaphysics do not serve for the *description of states of affairs*, neither existing ones (in that case they would be true statements) nor non-existing ones (in that case they would be at least false statements). They serve for the expressions of the general attitude of a person towards life [*Lebenseinstellung, Lebensgefühl*].

> ('Überwindung der Metaphysik' [1932], trans. in A. J. Ayer (ed.), *Logical Positivism* (New York: Macmillan, 1966), pp. 78–9)

15 Buddhism also makes suffering central to its conception of human existence, though in a different way, since (as I understand it) suffering is not regarded as redemptive, but as something ultimately to be escaped, through enlightenment.

16 L. Kolakowski, *Religion* (New York: Oxford University Press,1982), Ch. 5, p.188. He aptly adds that 'this was what made Christianity so hateful to Nietzsche'.

17 See, for example, Aristotle, *Nicomachean Ethics* [C. 325BC], Books II, III, X; the references to autonomous rational agency are all pervasive, those to human fragility few (see however Book I, Ch. 10). I take my cue here from Alasdair MacIntyre:

> [In Western moral philosophy] moral agents . . . are . . . presented as though they were continuously rational, healthy and untroubled . . . Aristotle . . . anticipated . . . a great many . . . in importing into moral philosophy the standpoint of those who have taken themselves to be self-sufficiently superior and of those who take their standards from those who take themselves to be self-sufficiently superior.

> (*Dependent Rational Animals* (London: Duckworth, 1999), pp. 1–2, 7)

18 Psalm 89 (Book of Common Prayer version).

19 Augustine, *De Civitate* [420], ix 13.

20 Martin Heidegger, *Sein und Zeit* [1927]; 8th edn (Tubigen: Niemeyer, 1957), p. 189. Trans. as *Being and Time*, J. Macquarrie and E. Robinson (New York: Harper and Row, 1962), p. 233.

21 Rainer Maria Rilke, *Duineser Elegien* [1922], I (trans. J. Cottingham). There is a German text with English translation in J.B. Leishman and S. Spender (eds), *Rainer Maria Rilke: Duino Elegies* (London: Hogarth Press, 1939; 4th edn 1968).

22 Pierre Hadot, *Philosophy as a Way of Life* (Oxford, UK and Cambridge, USA Blackwell, 1995), pp. 103–4. Originally published as *Exercises spirituels et philosophie antique* (Paris: Etudes Augustiniennes, 1987).

23 'I have come that they might have life, and that they might have it more abundantly' (John 10:10). For the images of the birds of the air and the lilies of the field, see Matthew 6: 25–9.

24 Isaiah 55:1.

25 Cited in Z. Sardar, *Muhammad for Beginners* (Cambridge: Icon, 1994), p. 60.

26 See P. Harvey, *An Introduction to Buddhist Ethics* (Cambridge: Cambridge University Press, 2000), Ch. 1.

27 Aristotle, *Nicomachean Ethics*, Book IV, Ch. 3. Compare J. Cottingham, *Philosophy and the Good Life* (Cambridge: Cambridge University Press, 1998), Ch.3, §6; and 'Partiality and the Virtues' in R. Crisp (ed.), *How Should One Live?* (Oxford: Clarendon, 1996), Ch.4,§1.

28 Compare Aristotle's discussion of the misfortunes that beset King Priam at the end of his life: *Nicomachean Ethics*, Book I, Ch. 10.

29 See Peter Singer, 'Famine, Affluence and Morality' [1972], repr. in J. Cottingham (ed.), *Western Philosophy* (Oxford: Blackwell, 1996), pp. 461ff.

30 Aldous Huxley, *Brave New World* (London: Chatto & Windus, 1932).

31 See John Keats, 'Ode to Melancholy' [1820].

32 T. S. Eliot, 'Burnt Norton' (1935), II. In his *Four Quartets* [1943].

33 A. N. Wilson, *Love Unknown* [1986] (Harmondsworth: Penguin, 1987), pp. 59–60.

34 George Eliot, *Adam Bede* [1859], Ch. 18.

35 Eliot, *Adam Bede*, Ch. 17.

36 See M. Foucault, Seminar at the Collège de France of 6 January 1982. Published as 'Subjectivité et vérité' in Y. C. Zarka (ed.), *Cités* (Vendôme: Presses Universitaires de France), vol. 2 (March 2000), pp. 155–6.

37 Colossians 3: 11; Galatians 3: 28.

38 See for example the fierce phrasing of Galatians 5:12.

39 Letter to the Romans [c. AD50] 2: 29, 'peritome kardias en pneumati, ou grammati' ('circumcision of the heart, in the spirit and not in the letter').

40 At Caesarea Philippi; Mark 8: 27.

41 See G. Vermes, *Jesus the Jew: A Historian's Reading of the Gospels* (London: Collins, 1973), Ch. 8.

42 The Arian controversy of the fourth century involved fierce debates between 'anomeans', 'homoeans' and 'homoiousians' on whether the Son was created by the Father, and whether his substance was of the same or a different nature, or of some complex intermediate status.

43 The Son is the 'image (*eikon*) of the invisible God' (Colossians 1:15).

> Christ is the image of God . . . For God, who commanded to the light to shine out of darkness, hath shined in our hearts, to give the light of the knowledge of the glory of God in the face of Jesus Christ.
>
> (2 Corinthians 4: 4–6)

44 See the references to Wittgenstein and others in note 14 above.

45 There would, of course, be complex metaphysical accounts to be given of the way in which the divine nature functions as the 'source' for each of these three respective features. But the underlying idea is a relatively simple one: that God's creative acts give rise to the objective realities of the cosmos that our beliefs conform to if true; and further, that it is from the way in which the cosmos (including our own human nature) is ordered that there arise those objective values which our moral and aesthetic judgements, if sound, reflect. (Those suspicious of the very idea of objectivity in the domain of value judgements might profitably consult Paul Bloomfield, *Moral Reality* (Oxford: Oxford University Press, 2001), where the context of the debate is, however, an entirely secular one.)

46 It is in this sense that Kantian ethics, with its stress on autonomy and freedom and the purity or 'holiness' of the will, has more in common with religious frameworks for living than do utilitarian or virtue ethics, which stress, respectively, the maximization of happiness and the development of human excellences. For the 'holy' will, see *Groundwork*

[*Grundlegung zur Metaphysik der Sitten*, 1785], trans. in H. J. Paton, *The Moral Law* (London: Hutchinson, 1948), p. 101.

47 Aristotle, *Nicomachean Ethics*, Book III, Ch. 5.

48 W. K. Clifford, *The Ethics of Belief* (1879). Repr. in B. Davies, *Philosophy of Religion* (Oxford: Oxford University Press, 2000).

49 *Critique of Pure Reason* [*Kritik der Reinen Vernunft*, 1781, 1787], B xxx (trans. N. Kemp Smith (New York: St Martin's Press, 1965), p. 29. Kant's term (prefiguring Hegel) is *aufheben*: implying not so much that faith 'denies' knowledge in order to make room for faith (as Kemp Smith's translation misleadingly has it), as that it transcends it. Cf. H. Kaygill, *A Kant Dictionary* (Oxford: Blackwell, 1995), s.v. 'faith'.

50 See Michael McGhee, *Transformations of Mind* (Cambridge: Cambridge University Press, 1999), p. 167.

51 From the hymn *Pange lingua* [1260]. The relevant line is 'prastet fides supplementum sensuum defectui' ('let faith make up for the deficiency of the senses'). Aquinas does not maintain what is sometimes called a 'fideist' position, that faith *substitutes* for reason; the two, rather, are complementary, with an 'ascent' to God via natural knowledge (supported by reason) going hand in hand with a 'descent' from God, through grace, of supernatural knowledge (based on faith). *Summa contra Gentiles* [1160], trans. in A. C. Pegis (Notre Dame, Ill.: Notre Dame University Press, 1975), p. 39.

52 Blaise Pascal, *Pensées* ('Thoughts') [1670], ed. L. Lafuma (Paris: Seuil, 1962), no. 418 (trans. J. Cottingham, emphasis supplied). For an excellent discussion of the whole passage, see Ward E. Jones, 'Religious Conversion, Self-Deception and Pascal's Wager', *Journal of the History of Philosophy*, 36:2 (April 1998).

53 Aristotle, *Nicomachean Ethics*, Book II, Ch. 1.

54 L. Wittgenstein, *Lectures and Conversations on Aesthetics, Psychology and Religious Belief* [1936–8], ed. C. Barrett (Oxford: Blackwell, 1966), pp. 58–9.

55 Wittgenstein, *Culture and Value*, ed. G. H. von Wright, trans. P. Winch (Oxford: Blackwell, 1980), pp. 64, 85.

56 H.-J. Glock, *A Wittgenstein Dictionary* (Oxford: Blackwell, 1996), p. 323.

57 L. R. Kass, *The Hungry Soul: Eating and the Perfecting of Our Nature* (New York: Macmillan, 1994), pp. 228–31.

58 'Bedürfnis der Vernunft', *Critique of Practical Reason* [*Kritik der Practischen Vernunft*, 1788], Part I, Book II, Ch. 2, §viii, in *Kant's gesammelte Schriften*,

(Akademie edn, Berlin: Reimer/De Gruyter, 1900), 5: 141 (trans. T. K. Abbott, *Critique of Practical Reason* (London: Longmans, 1873; 6th edn 1909), pp. 240ff. For Kant's views on faith, see Alan Wood, 'Rational theology, moral faith and reason', in P. Guyer (ed.), *The Cambridge Companion to Kant* (Cambridge: Cambridge University Press, 1992), Ch. 13, especially pp. 404–5.

59 'Wovon man nicht sprechen kann, darüber muß man schweigen' (Wittgenstein, *Tractatus Logico-Philosophicus* [1921], proposition 7). Making the alteration suggested, we might say instead '. . . davon muß man handeln'.

60 Thomas Traherne, 'The Third Century' [c. 1670], §3, in *Centuries, Poems and Thanksgivings*, ed. H. M. Margoliouth (Oxford: Oxford University Press, 1958), vol. 1, p. 111. Quoted in J. V. Taylor, *The Christlike God* (London: SCM, 1992), p. 33.

61 Bernard of Clairvaux, *Sermon on the Song of Songs*, v. 14 Quoted in Sarah Coakley, 'Visions of the Self in Late Medieval Christianity', in M. McGhee (ed.), *Philosophy and the Spiritual Life* (Cambridge: Cambridge University Press, 1992), p.95.

62 See, for example, T. Nagel, *Mortal Questions* (Cambridge: Cambridge University Press, 1979), Ch. 12. Contrast F. Jackson, 'The Primary Quality View of Colour' in his *From Metaphysics to Ethics* (Oxford: Clarendon, 1998), Ch. 4.

63 William Blake, 'Auguries of Innocence', from the Pickering manuscript [c. 1803].

64   There are in our existence spots of time
           Which with distinct pre-eminence retain
           A renovating virtue, whence . . . our minds
           Are nourished and invisibly repaired.

       (William Wordsworth, *Prelude* [1799; rev. 1805], Book XI)

65 ''Tis a common observation, that the mind has a great propensity to spread itself on external objects . . .', David Hume, *A Treatise of Human Nature* [1739–40], Book I, part 3, section xiv (ed. L. A. Selby-Bigge, rev. P. H. Nidditch, 2nd edn, (Oxford: Clarendon, 1978), p. 167).

66 Immanuel Kant, *Critique of Practical Reason*, Conclusion, (Akademie edn. 5: 161 (trans. T. K. Abbott, p. 260)).

67 William Shakespeare, *Macbeth* [c. 1605] Act 5, scene 5.

68 I take the phrase from the title of a thesis by Jane Waterworth, *Living in the Light of Hope* (Umea University, 2001), though I should add that Waterworth aims to provide an entirely secularised account of this notion.

Index